More Than Human

More Than Human

[EMBRACING THE PROMISE OF BIOLOGICAL ENHANCEMENT]

Ramez Naam

Broadway Books

New York   London   Toronto   Sydney   Auckland

Broadway Books titles may be purchased for business or promotional use or for special
sales. For information, please write to: Special Markets Department,
Random House, Inc., 1745 Broadway, New York, NY 10019.

PRINTED IN THE UNITED STATES OF AMERICA

BROADWAY BOOKS and its logo, a letter B bisected on the diagonal,
are trademarks of Random House, Inc.

Visit our website at www.broadwaybooks.com

First edition published 2005.

Book design by Chris Welch
Illustrations by the author

Library of Congress Cataloging-in-Publication Data
Naam, Ramez.
More than human : embracing the promise of biological enhancement / Ramez Naam.
p. cm.
1. Medical genetics—Popular works. 2. Genetic engineering—Popular works.
3. Human reproductive technology—Popular works. I. Title.
RB155.N27 2005
616'.042—dc22
2004054507

ISBN 0-7679-1843-6

1   3   5   7   9   10   8   6   4   2

Dedicated to Mom and Dad and Mary and Natalia,
for all your love

Contents

More
Than
Human

Healing
and
Enhancing

SOUTHERN ILLINOIS UNIVERSITY Medical School rises above the rolling cornfields that surround the state capital of Springfield, three hours south of Chicago. Prairie dogs, field mice, and other animals inhabit the country-side. In a lab in a nondescript building on the campus, a thin, bald-ing, smiling man named Andrzej Bartke has been investigating the causes of human aging. His goal is to learn how to delay the onset of heart disease, cancer, frailty, and other problems that come with age—to increase the human health span, as he calls it.

Bartke, smiling, holds two mice in his cupped hands. One, a close cousin of the mice that scamper around in the fields, is of a breed whose members live for two and a half years on average be-fore succumbing to old age. The other mouse is a breed that has been genetically engineered to live longer. The oldest of its kind died just a week short of its fifth birthday, twice as long as a geneti-cally normal mouse—it's as if a human lived to be two hundred. The difference between the breeds is just a change to a single gene, a gene that exists in everything from yeast to mice to men and women. In seeking to prevent disease, Bartke and his colleagues have opened the possibility of doubling the human life span.

The same basic story has repeated itself dozens of times in the past decade. In the new field of biotechnology, researchers looking for therapies, ways to heal the sick or injured or preserve the health of the elderly, have stumbled on techniques that might enhance human abilities. At Princeton University researchers looking for ways to stave off Alzheimer's disease have created mice that can learn things more than twice as fast as their genetically normal peers. Their research is pointing to techniques that could improve human learning as well.

In a lab at the University of Pennsylvania, researchers trying to find a cure for muscular dystrophy have produced mice that are superstrong. They can climb things and lift things that their normal siblings can't. At some point in their lives, each has been injected with extra copies of a gene that influences muscle building. The same gene plays a role in human muscle building.

At Duke University, scientists looking for ways to help amputees and paralytics have implanted electrodes in the brains of a group of monkeys. The monkeys can move mechanical arms *just by thinking about it,* as if those robots were parts of their bodies.

In Lisbon, Portugal, there's a group of blind men and women who can now see. In place of eyeglasses they wear cameras connected to electrodes implanted in the visual area of their brains. Some of them had been blind for twenty years or more before the surgery. The technology that has given them sight could beam images from one person's mind to another's.

The scientists conducting these studies have a common goal: to heal sick and injured men and women. Along the way, they're discovering that to heal our minds and bodies we must understand them, and when we understand we gain the power to improve. That power could give us better learning and memory, better muscles, and longer lives. By unraveling how our minds and bodies function, biotechnology could give us the power to sculpt any aspect of ourselves—how we think, how we feel, how we look, how we communicate with one another.

Not everyone is convinced that this power to alter ourselves is a benefit to society. There are those who oppose too much tinkering with the human mind and body. These concerns have been voiced across the political spectrum. In 2000, President George W. Bush created the President's Council on Bioethics to advise him on issues of biotechnology. To head the council, he appointed Leon Kass, a University of Chicago professor of bioethics and a political conservative with a twenty-five-year history of opposing infertility treatments, cosmetic surgery, organ transplantation, and other technologies that, in his view, violate the natural order of things. Under Kass's direction, the council has released report after report condemning the use of new biotechnologies to alter the human mind or body.

The council's 2004 report, "Beyond Therapy," argues that genetic and reproductive technologies undermine the value of life and disrupt the natural relationship between parents and children; that slowing human aging would cause social stagnation as the old cling stubbornly to power; that instant performance enhancers would reduce human drive and hard work; that technologies for sculpting the human mind would threaten our sense of identity; that techniques to enhance human abilities might widen the gap between rich and poor; that the safety of these techniques cannot be assured; and that they could be used by the powerful to coerce, control, or gain an advantage over the weak.

Ultimately the council argues that we ought to revere our given, natural state, that to seek to improve on what we have is hubris, and that biotechnological alteration of our minds and bodies threatens our unique human dignity.

In his book *Our Posthuman Future,* bioethics council member Francis Fukuyama echoes this sentiment. "There are good prudential reasons to defer to the natural order of things and not to think that human beings can easily improve on it."

Given these concerns, how should society react? Fukuyama believes that "there are certain things that should be banned outright. One of them is reproductive cloning. . . . If we get used to cloning in the near term, it will be much harder to oppose germline [genetic] engineering for enhancement purposes in the future." Fukuyama would like to restrict more than just technologies for engineering genes, arguing that governments need to "draw red lines" around technologies in general, "to distinguish between therapy and enhancement, directing research toward the former while putting restrictions on the latter."

George Annas, a liberal philosopher and bioethicist, has called genetic engineering a "crime against humanity" and has called for an international treaty prohibiting it. William Kristol, the publisher and editor of the conservative *Weekly Standard,* agrees: "A ban on cloning would only be a first step down the road of responsibility and self-government—but it would be an important step."

Environmentalist and author Bill McKibben goes further, calling for a halt on even the basic scientific research that might lead to techniques to enhance human abilities. In his book *Enough,* he writes, "We need to do an unlikely thing: we need to survey the world we now inhabit and proclaim it good. Good enough. . . . Enough intelligence. Enough capability. Enough."

In short, a chorus of voices now argue that we should strive to preserve the status quo, that we should opt for stability over change, for the known over the unknown. To prevent unwanted changes, Fukuyama says, "We should use the power of the state" to restrict access to technologies that might undermine our current notions of humanity, that might allow individuals to surpass the mental and physical limitations we now know.

This book is based on a very different premise: rather than fearing change, we ought to embrace it, rather than prohibiting the exploration of new technologies, society ought to focus on spreading the power to alter our own minds and bodies to as many people as possible. Rather than imposing a rigid view of what it means to be human on humanity, we ought to trust billions of individuals and families to decide that for themselves. In making that case, I'm going to show you four things.

FIRST, THERE ARE good pragmatic reasons to embrace human enhancement. Scientists cannot draw a clear line between healing and enhancing, for they're integrally related. Promising research on curing Alzheimer's disease, on reducing the incidence of heart disease and cancer, on restoring sight to the blind and motion to the paralyzed is the very same research that could lead to keeping us young, improving our memories, wiring our minds together, or enhancing ourselves in other ways. We cannot stop research into enhancing ourselves without also halting research focused on healing the sick and injured. If we are to ban such improvements to our quality of life, we had better have strong evidence that the research poses a greater threat to society than the medical benefits it brings.

Even beyond curing disease, the benefits to be won from biotechnology are concrete and measurable. Keeping people young longer would slow the rise in worldwide health spending and avoid the demographic crunch of an aging population around the world. Improving human memory, attention, and communication abilities would increase productivity, which in turn would lead to new scientific discoveries and faster innovation, economic growth, and scientific breakthroughs we can't anticipate today.

SECOND, BANNING BIOTECHNOLOGICAL research wouldn't solve any problems—it would only make matters worse. Steven Hyman, a former director of the National Institute of Mental Health, told the President's Council on Bioethics, "Regulation that flies in the face of reality . . . decreases respect for regulation and fails. . . . If we can find medications which will enhance our performance, lengthen our life, decrease the stress . . . no regulation in the world would keep them from general use."

Prohibition of enhancement technologies wouldn't stop people from seeking to use them any more than the War on Drugs has stopped people from seeking recreational chemicals. A ban would make enhancement procedures more expensive by creating a black market with an artificially high price, further separating the rich and poor. It would add risk for those who sought enhancement, by removing any possibility of safety regulations. Furthermore, a ban would hinder scientific progress, as long-term studies of the effects of these technologies would become impossible. Biotechnology would be driven underground, and the public would be left in the dark as to the research being conducted.

THIRD, THE DEBATE over human enhancement is at heart a debate over human freedom. Should individuals and families have the right to alter their own minds and bodies, or should that power be held by the state?

In a democratic society, it's every man and woman who should determine such things, and not the state. Western democracies are founded on the principle that governments exist to protect the freedoms of individuals, or so the Declaration of Independence claims:

> We hold these truths to be self-evident, that all men are created equal, that they are endowed by their Creator with certain un-alienable Rights, that among these are Life, Liberty and the pur-suit of Happiness. —That to secure these rights, Governments are instituted among Men. . . .

Yet beneath the arguments of bioconservatives, advocates of the biological status quo, is the unspoken assumption that a small cadre of elites—legislators, regulators, and professional ethicists—know what's best, that they can make better decisions on behalf of millions or billions of people than those individuals could make for themselves.

For example, in an essay on the perils of lengthening human life, President Bush's top ethicist, Leon Kass, has written, "The finitude of life is a blessing for every individual, whether he knows it or not." Is this really for Professor Kass to decide?

Governments are instituted to *secure* individual rights, not to re-strict them. This is an issue of both principle and good gover-nance: recent history suggests that nations that embrace the freedoms of individuals thrive, whereas nations that limit individ-ual rights fall behind and fail, often in monstrous ways.

In 1992, before the publication of his antibiotech treatise, *Our Posthuman Future,* Francis Fukuyama wrote a book titled *The End of History.* Fukuyama argued, as other historians have, that the twentieth century saw the victory of liberal democracy over totalitarian forms of government.

At the start of the twentieth century, political systems like Communism were formed around the belief that government could make better decisions for people than the people could for themselves. Democracies were founded more or less on the opposite principle—that individuals making decisions for themselves would produce, on the whole, better results for everyone than central control.

At the end of the twentieth century, Fukuyama wrote, it was clear that liberal democracy had won. There could no longer be any doubt that free societies, societies where individuals made most of the decisions for themselves, could outcompete centrally controlled societies economically, scientifically, and technologically. Just as important, there could no longer be any doubt that individuals living in free societies were happier and better off than those living anywhere else in the world.

The triumph of democracy has its roots in two simple phenomena. Individuals generally want to improve their situation. They want to live longer, remain in good health, increase their capabilities, expand their options, increase the well-being of their children, and so on. Given the choice, they'll work in this direction. In addition, millions of individuals weighing costs and benefits have a greater collective intelligence, better collective judgment, than a small number of centralized regulators and controllers. People aren't perfect decision makers by any means—individuals make a tremendous number of mistakes—but on the whole the distributed decision-making power of the masses is more effective at moving society toward greater welfare than the decisions made by a small elite.

Those two simple facts explain why a market economy driven by the often rather shallow desires of consumers nonetheless out-competes and out-innovates a centrally controlled economy. They explain why political systems based on votes cast by a public with widely varying views are nonetheless more stable and more effective at preserving the rights of citizens than monarchies or dictatorships.

If the lessons of the twentieth century hold true into the twenty-first, the nations that attempt to maintain control over the minds and bodies of their people will fall behind, while nations that embrace human freedom to alter our own minds and bodies will thrive.

FINALLY, FAR FROM being unnatural, the drive to alter and improve on ourselves is a fundamental part of who we humans are. As a species we've always looked for ways to be faster, stronger, and smarter and to live longer. Many past enhancements that we now take for granted—from blood transfusions to vaccinations to birth control—were called unnatural or immoral when they were first introduced. Yet over time we've become accustomed to these new levels of control over our minds and bodies, and have used them for the betterment of ourselves, our families, and our world.

New discoveries have at times caused upheavals in our notion of who we are and what our lives mean. Galileo's discovery that the earth is not the center of the universe challenged the seventeenth-century conception of our place in the cosmos. Darwin's theory of evolution challenged the nineteenth-century conception of humanity's place in nature. Now science is challenging the idea that our mental and physical forms are fixed. We can change them if we choose.

Yet our lives today are no less meaningful and no less challenging than they were centuries ago. If anything, greater knowledge and power over ourselves and our world has enriched us. Technologies that have increased our abilities—telephones, cars, books, and medicine—have increased our opportunities to connect with one another, to learn about our world, to contribute to the rest of humanity, and thus to lead more meaningful lives.

The only reason we live lives of such great comfort and potential today is that throughout history, there have been men and women who've refused to accept the natural order of things. From the mastery of fire, to agriculture, to written language, to eyeglasses, to antibiotics, previous generations constantly improved their world, their minds, and their bodies in search of a better life for themselves and their families. The benefits of that constant quest to improve the human situation have been passed down to us. Men and women striving for better lives probably never imagined the impact their efforts would have decades, centuries, or millennia down the line. Whatever efforts we make to improve our own lives and the lives of our children will benefit future generations in ways that we ourselves can't anticipate.

In the end, this search for ways to enhance ourselves is a natural part of being human. The urge to transform ourselves has been a force in history as far back as we can see. It's been selected for by millions of years of evolution. It's wired deep in our genes—a natural outgrowth of our human intelligence, curiosity, and drive. To turn our backs on this power would be to turn our backs on our true nature. Embracing our quest to understand and improve on ourselves doesn't call into question our humanity—it reaffirms it.

CHAPTER 1

Choosing Our Bodies

N 1989, RAJ and Van DeSilva were desperate. Their daughter Ashanti, just four, was dying. She was born with a crippled immune system, a consequence of a problem in her genes.

Every human being has around thirty thousand genes. In fact, we have two copies of each of those genes—one inherited from our mother, the other from our father. Our genes tell our cells what proteins to make, and when.

Each protein is a tiny molecular machine. Every cell in your body is built out of millions of these little machines, working together in precise ways. Proteins break down food, ferry energy to the right places, and form scaffoldings that maintain cell health and structure. Some proteins synthesize messenger molecules to pass signals in the brain, and other proteins form receptors to receive those signals. Even the machines inside each of your cells that build new proteins—called ribosomes—are themselves made up of other proteins.

Ashanti DeSilva inherited two broken copies of the gene that contains the instructions for manufacturing a protein called adenoside deaminase (ADA). If she had had just one broken copy, she would have been fine. The other copy of the gene would have made up the difference. With two broken copies, her body didn't have the right instructions to manufacture ADA at all.

ADA plays a crucial role in our resistance to disease. Without it, special white blood cells called T cells die off. Without T cells, ADA-deficient children are wide open to the attacks of viruses and bacteria. These children have what's called severe combined immune deficiency (SCID) disorder, more commonly known as bubble boy disease.

To a person with a weak immune system, the outside world is threatening. Everyone you touch, share a glass with, or share the same air with is a potential source of dangerous pathogens. Lacking the ability to defend herself, Ashanti was largely confined to her home.

The standard treatment for ADA deficiency is frequent injections of PEG-ADA, a synthetic form of the ADA enzyme. PEG-ADA can mean the difference between life and death for an ADA-deficient child. Unfortunately, although it usually produces a rapid improvement when first used, children tend to respond less and less to the drug each time they receive a dose. Ashanti DeSilva started receiving PEG-ADA injections at the age of two, and initially she responded well. Her T-cell count rose sharply and she developed some resistance to disease. But by the age of four, she was slipping away, no longer responding strongly to her injections. If she was to live, she'd need something more than PEG-ADA. The only other option at the time, a bone-marrow transplant, was ruled out by the lack of matching donors.

In early 1990, while Ashanti's parents were searching frantically for help, French Anderson, a geneticist at the National Institutes of Health, was seeking permission to perform the first gene-therapy trials on humans. Anderson, an intense fifth-degree black-belt in tae kwon do and respected researcher in the field of genetics, wanted to show that he could treat genetic diseases caused by faulty copies of genes by inserting new, working copies of the same gene.

Scientists had already shown that it was possible to insert new genes into plants and animals. Genetic engineering got its start in 1972, when geneticists Stanley Cohen and Herbert Boyer first met at a scientific conference in Hawaii on plasmids, small circular loops of extra chromosomal DNA in which bacteria carry their genes. Cohen, then a professor at Stanford, had been working on ways to insert new plasmids into bacteria. Researchers in Boyer's lab at the University of California in San Francisco had recently discovered restriction enzymes, molecular tools that could be used to slice and dice DNA at specific points.

Over hot pastrami and corned-beef sandwiches, the two Californian researchers concluded that their technologies complemented one another. Boyer's restriction enzymes could isolate specific genes, and Cohen's techniques could then deliver them to bacteria. Using both techniques researchers could alter the genes of bacteria. In 1973, just four months after meeting each other, Cohen and Boyer inserted a new gene into the *Escherichia coli* bacterium (a regular resident of the human intestine).

For the first time, humans were tinkering directly with the genes of another species. The field of genetic engineering was born. Boyer would go on to found Genentech, the world's first biotechnology company. Cohen would go on to win the Nobel Prize in 1986 for his work on cell growth factors.

Building on Cohen and Boyer's work with bacteria, hundreds of scientists went on to find ways to insert new genes into plants and animals. The hard work of genetically engineering these higher organisms lies in getting the new gene into the cells. To do this, one needs a gene vector—a way to get the gene to the right place. Most researchers use gene vectors provided by nature: viruses. In some ways, viruses are an ideal tool for ferrying genes into a cell, because penetrating cell walls is already one of their main abilities. Viruses are cellular parasites. Unlike plant or animal cells, or even bacteria, viruses can't reproduce themselves. Instead, they penetrate cells and implant their viral genes; these genes then instruct the cell to make more of the virus, one protein at a time.

Early genetic engineers realized that they could use viruses to deliver whatever genes they wanted. Instead of delivering the genes to create more virus, a virus could be modified to deliver a different gene chosen by a scientist. Modified viruses were pressed into service as genetic "trucks," carrying a payload of genes loaded onto them by researchers; these viruses don't spread from cell to cell, because they don't carry the genes necessary for the cell to make new copies of the virus.

By the late 1980s, researchers had used this technique to alter the genes of dozens of species of plants and animals—tobacco plants that glow, tomatoes that could survive freezing, corn resistant to pesticides. French Anderson and his colleagues reasoned that one could do the same in a human being. Given a patient who lacked a gene crucial to health, one ought to be able to give that person copies of the missing gene. This is what Anderson proposed to do for Ashanti.

Starting in June of 1988, Anderson's proposed clinical protocols, or treatment plans, went through intense scrutiny and generated more than a little hostility. His first protocol was reviewed by both the National Institutes of Health (NIH) and the Food and Drug Administration (FDA). Over a period of seven months, seven regulatory committees conducted fifteen meetings and twenty hours of public hearings to assess the proposal.

In early 1990, Anderson and his collaborators received the final approval from the NIH's Recombinant DNA Advisory Committee and had cleared all legal hurdles. By spring, they had identified Ashanti as a potential patient. Would her parents consent to an experimental treatment? Of course there were risks to the therapy, yet without it Ashanti would face a life of seclusion and probably death in the next few years. Given these odds, her parents opted to try the therapy. As Raj DeSilva told the *Houston Chronicle,* "What choice did we have?"

Ashanti and her parents flew to the NIH Clinical Center at Bethesda, Maryland. There, over the course of twelve days, Anderson and his colleagues Michael Blaese and Kenneth Culver slowly extracted some of Ashanti's blood cells. Safely outside the body, the cells had new, working copies of the ADA gene inserted into them by a hollowed-out virus. Finally, starting on the afternoon of September 14, Culver injected the cells back into Ashanti's body.

The gene therapy had roughly the same goal as a bone-marrow transplant—to give Ashanti a supply of her own cells that could produce ADA. Unlike a bone-marrow transplant, gene therapy carries no risk of rejection. The cells Culver injected back into Ashanti's bloodstream were her own, so her body recognized them as such.

The impact of the gene therapy on Ashanti was striking. Within six months, her T-cell count rose to normal levels. Over the next two years, her health continued to improve, allowing her to enroll in school, venture out of the house, and lead a fairly normal childhood.

Ashanti is not completely cured—she still takes a low dose of PEG-ADA. Normally the dose size would increase with the patient's age, but her doses have remained fixed at her four-year-old level. It's possible that she could be taken off the PEG-ADA therapy entirely, but her doctors don't think it's yet worth the risk. The fact that she's alive today—let alone healthy and active—is due to her gene therapy, and also helps prove a crucial point: genes can be inserted into humans to cure genetic diseases.

From Healing to Enhancing

After Ashanti's treatment, the field of gene therapy blossomed. Since 1990, hundreds of labs have begun experimenting with gene therapy as a technique to cure disease, and more than five hundred human trials involving over four thousand patients have been launched. Researchers have shown that it may be possible to use gene therapy to cure diabetes, sickle-cell anemia, several kinds of cancer, Huntington's disease and even to open blocked arteries.

While the goal of gene therapy researchers is to cure disease, gene therapy could also be used to boost human athletic performance. In many cases, the same research that is focused on saving lives has also shown that it can enhance the abilities of animals, with the suggestion that it could enhance men and women as well.

Consider the use of gene therapy to combat anemia. Circulating through your veins are trillions of red blood cells. Pumped by your heart, they serve to deliver oxygen from the lungs to the rest of your tissues, and carry carbon dioxide from the tissues back out to the lungs and out of the body. Without enough red blood cells, you can't function. Your muscles can't get enough oxygen to produce force, and your brain can't get enough oxygen to think clearly. Anemia is the name of the condition of insufficient red blood cells. Hundreds of thousands of people worldwide live with anemia, and with the lethargy and weakness that are its symptoms. In the United States, at least eighty-five thousand patients are severely anemic as a result of kidney failure. Another fifty thousand AIDS patients are anemic due to side effects of the HIV drug AZT.

In 1985, researchers at Amgen, a biotech company based in Thousand Oaks, California, looking for a way to treat anemia isolated the gene responsible for producing the growth hormone erythropoietin (EPO). Your kidneys produce EPO in response to low levels of oxygen in the blood. EPO in turn causes your body to produce more red blood cells. For a patient whose kidneys have failed, injections of Amgen's synthetic EPO can take up some of the slack. The drug is a lifesaver, so popular that the worldwide market for it is as high as $5 billion per year, and therein lies the problem: the cost of therapy is prohibitive. Three injections of EPO a week is a standard treatment, and patients who need this kind of therapy end up paying $7,000 to $9,000 a year. In poor countries struggling even to pay for HIV drugs like AZT, the added burden of paying for EPO to offset the side effects just isn't feasible.

What if there was another way? What if the body could be instructed to produce more EPO on its own, to make up for that lost to kidney failure or AZT? That's the question University of Chicago professor Jeffrey Leiden asked himself in the mid-1990s. In 1997, Leiden and his colleagues performed the first animal study of EPO gene therapy, injecting lab monkeys and mice with a virus carrying an extra copy of the EPO gene. The virus penetrated a tiny proportion of the cells in the mice and monkeys and unloaded the gene copies in them. The cells began to produce extra EPO, causing the animals' bodies to create more red blood cells. In principle, this was no different from injecting extra copies of the ADA gene into Ashanti, except in this case the animals already had two working copies of the EPO gene. The one being inserted into some of their cells was a third copy; if the experiment worked, the animals' levels of EPO production would be boosted beyond the norm for their species.

That's just what happened. After just a single injection, the animals began producing more EPO, and their red-blood-cell counts soared. The mice went from a hematocrit of 49 percent (meaning that 49 percent of their blood volume was red blood cells) to 81 percent. The monkeys went from 40 percent to 70 percent. At least two other biotech companies, Chiron and Ariad Gene Therapies, have produced similar results in baboons and monkeys, respectively.

The increase in red-blood-cell count is impressive, but the real advantage of gene therapy is in the long-lasting effects. Doctors can produce an increase in red-blood-cell production in patients with injections of EPO itself—but the EPO injections have to be repeated three times a week. EPO gene therapy, on the other hand, could be administered just every few months, or even just once for the patient's entire lifetime.

The research bears this out. In Leiden's original experiment, the mice each received just one shot, but showed higher red-blood-cell counts for a year. In the monkeys, the effects lasted for twelve weeks. The monkeys in the Ariad trial, which went through gene therapy more than four years ago, still show higher red-blood-cell counts today.

This is a key difference between drug therapy and gene therapy. Drugs sent into the body have an effect for a while, but eventually are broken up or passed out. Gene therapy, on the other hand, gives the body the ability to manufacture the needed protein or enzyme or other chemical itself. The new genes can last for a few weeks or can become a permanent part of the patient's genome.

The duration of the effect depends on the kind of gene vector used and where it delivers its payload of DNA. Almost all of the DNA you carry is located on twenty-three pairs of chromosomes that are inside the nuclei of your cells. The nucleus forms a protective barrier that shields your chromosomes from damage. It also contains sophisticated DNA repair mechanisms that patch up most of the damage that does occur.

Insertional gene vectors penetrate all the way into the nucleus of the cell and splice the genes they carry into the chromosomes. From that point on, the new genes get all the benefits your other genes enjoy. The new genes are shielded from most of the damage that can happen inside your cells. If the cell divides, the new genes get copied to the daughter cells, just like the rest of your DNA. Insertional vectors make more or less permanent changes to your genome.

Noninsertional vectors, on the other hand, don't make it into the nucleus of your cells. They don't splice the new genes they carry into your chromosomes. Instead, they deliver their payload of DNA and leave it floating around inside your cells. The new DNA still gets read by the cell. It still instructs the cell to make new proteins. But it doesn't get copied when the cell divides. Over time, it suffers from wear and tear, until eventually it breaks up, and its effects end.

The difference in durations among drugs, noninsertional vectors, and insertional vectors gives us choices. We can choose to make a temporary change with a drug, which will wear off in a few hours or days; a semipermanent change with noninsertional gene therapy, whose effects will last for weeks or months depending on the genes and the type of cell infected; or a permanent change by inserting new genes directly into your genome. Each of these three options is appropriate in certain situations. In the context of EPO, the idea of semipermanent or permanent change by means of gene therapy has definite advantages. It cuts down on the need for frequent injections, which means that the gene therapy approach can end up being much cheaper than the drug therapy approach.

Boosting Athletic Performance

If you've heard of EPO, you may be a sports fan. The Tour de France is the most famous and prestigious cycling race in the world. In 1998, while the race was ongoing, customs officials at the border between Belgium and France found more than four hundred doses of EPO in the car of Willy Voet, the massage therapist for Team Festina—one of the prime contenders in the race. Within days, the mostly Swiss and French team was disqualified.

Injections of EPO cause your body to produce more red blood cells, whether you're anemic or not. With more red blood cells, your body can carry more oxygen. With greater oxygen-carrying capacity, your endurance increases. If you're an elite athlete, injections of synthetic EPO can boost your performance in distance

running, swimming, and bicycling events by as much as 10 percent. In these sports, where victory is often snatched by a few frac-
tions of a second, that's an incredible advantage.

EPO isn't the first drug athletes have used to boost performance.
After the 1988 Olympics, sprinter Ben Johnson was stripped of his
gold medal after testing positive for anabolic steroids. Baseball
player and 1999 home-run record setter Mark McGwire has admit-
ted to using the steroid androstenedione. Other athletes have used
human growth hormone, methamphetamines, transfusions of additional human blood or hemoglobin, or synthetic blood additives
called PFCs (perflurorcarbons) to get a physical advantage.

Some of these athletic performance boosters have been
adopted by amateur athletes. For example, recreational weight-
lifters have been known to use steroids and human growth hormone. Yet the most effective drugs carry substantial health risks
that discourage widespread use. Anabolic steroids can cause heart
disease, high blood pressure, kidney problems, reduced sperm
count, and impotence in men, increased body hair and other mas-
culine features in women, and increased aggression in both. EPO
is generally safe at low doses, but too much can thicken the blood
to the point where the heart can no longer pump, leading to heart
attacks and death.

The inconvenience of these drugs also discourages widespread
use. EPO and many other performance-enhancing drugs require
multiple injections per week. The need for frequent doses means
that a user has to keep a supply of a usually illegal or prescription-
only substance on hand. Most amateurs looking for a small boost
to their tennis game aren't willing to subject themselves to frequent needle-pricks or the risks of regularly buying and possess-
ing a controlled substance.

Gene therapy changes the rules. EPO gene therapy would make it easier for athletes to cheat by reducing the risk of detection. Synthetic EPO is already hard to detect, because it's almost identical to the EPO produced by the body. Athletes using EPO aren't caught by blood tests. They're discovered only if they or their trainers are caught in possession of EPO. If an athlete could take a shot months before the event, instead of at the event itself, the risk of being caught in possession of the drug would drop dramatically.

The bigger impact of EPO gene therapy may be on recreational athletes. Few weekend hikers, bikers, or soccer players care enough about their performance to inject themselves three times a week. But what about an injection every few months? Or just one injection in a lifetime? Integrating gene vectors seems to make it possible to give a person an elevated red-blood-cell count and increased aerobic fitness and endurance *for life* with just a single injection.

Of course, EPO gene therapy would come with risks. There's the risk of gene therapy itself—the possibility that the newly inserted gene could disrupt the function of an existing gene, or that the body's immune system might respond. Yet there's a tremendous medical incentive to improve the safety and effectiveness of gene therapy, and a number of promising techniques have already shown success in human cells and live animals.

The more serious risk of adding an EPO gene is the risk of EPO itself. EPO has already been implicated in the deaths of quite a few athletes, including about twenty Dutch cyclists in the early 1990s. These athletes died because they took too much EPO. In pushing their red-blood-cell counts to the extreme, they thickened their blood to the point where their hearts could no longer pump it. Since then, the number of EPO deaths has declined as athletes have learned more about the drug. Aware of the risks, they can carefully moderate their dosage, cutting back a bit if their blood becomes too thick.

Biotech firms are also aware of these health risks and have come up with ingenious ways to throttle the amount of EPO added through gene therapy. These techniques depend on the *promoter* region of the gene, the part that tells the cell when it should actually create some of the protein the gene describes.

Some gene-therapy trials have the goal of producing as much of the protein the inserted gene describes as possible. In these cases, researchers use promoters that are "on" all the time. With EPO this isn't an option, since too much of the EPO protein can kill.

Instead of using an "always-on" promoter, researchers have come up with two different systems to control the activity of the gene. At Oxford Biomedica in the United Kingdom, researcher Katie Binley and her colleagues have developed a promoter tied to low oxygen levels in the blood. If EPO levels are high, blood-oxygen levels will remain high, and the gene won't be activated to produce more EPO. Another company, Ariad Gene Therapies, has created a promoter that's triggered by a modified version of the drug rapamycin. The modified rapamycin can be taken as a pill; once it's taken, it activates the extra EPO genes for about a month. That makes the gene-therapy-plus-pill approach quite a bit more convenient than thrice-weekly injections, and adding a degree of control over the EPO gene and the red-blood-cell count makes the therapy safer.

Gene-promoter techniques also illustrate some of the flexibility of gene therapy. We can control how long an effect lasts by selecting drug therapy or semipermanent or permanent gene therapy. And with selective gene promoters, we can also program *when* and *in which situation* a gene will become active. Ariad's promoter gene-therapy technique will soon enter human trials.

Designer Muscle

Every day, fourteen Americans are diagnosed with amyotrophic lateral sclerosis (ALS), better known as Lou Gehrig's disease. ALS is effectively a death sentence. On average, ALS patients live two to five years after diagnosis. Over those years, the nerve cells that control their muscles will die off. They'll progressively lose control of their bodies. The disease starts with twitching, cramping, and stiffness. Then the arms and legs become weak. Speech becomes slurred. Chewing and swallowing become difficult. Eventually ALS sufferers lose all voluntary muscle control. They become "locked in"—unable to move a muscle, unable to speak, but still perfectly aware of the world around them.

There's no known cure for ALS. The only approved treatment, a drug called riluzole, extends life by a few months at most. Recent animal studies, however, suggest that gene therapy could slow the progress of the disease and double the life expectancy of ALS patients.

In 2003, neuroscience professor Jeffrey Rothstein and his colleagues at Johns Hopkins University gave mice suffering the equivalent of ALS extra copies of a gene called IGF-1, for "insulin-like growth factor 1," a compound found in our bodies that, among other things, promotes muscle growth and repair.

ALS normally progresses so quickly in mice that researchers can't test therapies after the mice show symptoms; instead they have to test techniques on mice before symptoms manifest, and then see if tested mice do any better than mice that don't receive the treatment. The IGF-1 gene therapy worked differently. The researchers injected the mice after the symptoms first manifested. The mice responded amazingly well. They gained strength, and their nerve cells, the ones that normally die off in ALS, stayed healthy longer. Overall, they lived about twice as long as the mice who didn't receive the gene therapy.

Encouraged by these results, Johns Hopkins is now planning to conduct human trials of the therapy. If IGF-1 gene therapy protects humans against ALS, it will be the first truly effective weapon against the disease. Yet IGF-1 gene therapy may also find its way into other, nonmedical uses.

In 1998, University of Pennsylvania biologist Lee Sweeney and his team delivered extra copies of the IGF-1 gene into the muscles of healthy mice. Their goal was to see if IGF-1 could slow the age-related loss of muscle that occurs in both mice and humans. In Sweeney's first experiment, he injected the gene into eighteen-month-old mice, roughly the equivalent of fifty-five-year-old humans. He then measured the animals' strength and muscle mass when the mice were twenty-seven months old, the equivalent of eighty or older in humans.

To the researcher's surprise, the extra IGF-1 gene did more than slow down the age-related decline in strength—it practically halted it. The twenty-seven-month-old mice dosed with IGF-1 actually had a slight bit more muscle than they'd had at eighteen months. On some strength tests they showed only a tiny decline over the course of their lives, and on other strength tests they actually showed improvement.

Intrigued by his results, Sweeney decided to study the effects in young adult rats, about the equivalent of twenty-year-old humans. His team compared three groups of rats: (1) genetically normal rats put on an intense strength-training program for eight weeks and forced to climb up the sides of cages with weights tied to their bodies, (2) rats with an extra copy of IGF-1 that were completely sedentary; and (3), rats that got both the extra IGF-1 gene *and* the strength training.

The first surprise was that sedentary rats with IGF-1 gained as much strength, about 16 percent, as the genetically normal rats on the rigorous exercise plan. What's more, the sedentary IGF-1 group actually gained *more* muscle mass. But the group that received both the IGF-1 gene therapy and the exercise blew the other two groups away, increasing its strength by 30 percent in the same time period. Furthermore, two months after the study ended, the group that received only exercise had lost almost all of its strength gains, while the group that had received both IGF-1 and strength training retained more than half of its gains.

Other experiments have produced similar results with different genes. Another team led by Geoffrey Goldspink at the University of London gave mice extra copies of a variant of the IGF-1 gene called mechano growth factor, or MGF. The gene-dosed mice showed a 20 percent increase in muscle mass and a 25 percent increase in strength, without any exercise at all.

Se-Jin Lee and colleagues at Johns Hopkins University took a different approach. They bred a strain of mice that lacked a working copy of the gene for a hormone called myostatin. Myostatin normally suppresses the body's production of muscle. Lee's myostatin-deficient mice grew to have two to three times as much muscle as their genetically normal kin, and further study showed that they had 70 percent less body fat. Lee has gone on to find ways to reduce myostatin in adult mice. Knocking out myostatin in mice that suffer from the equivalent of muscular dystrophy helps them build muscle, leading researchers to investigate the potential of the therapy in humans suffering from the disease.

Gene therapy to boost muscle mass is likely to be even more popular than gene therapy to boost endurance. Unlike higher red-blood-cell counts, which can lead to heart attacks, there's no obvious health risk from large muscles.

Perhaps more important, the normal age-related decline in muscle mass and tone has now been medicalized—given a name and treated as a disease, not a "normal" occurrence. In 1991, two gerontologists, William Evans and Irwin Rosenberg, introduced the word *sarcopenia*, literally "vanishing flesh," to describe the phenomenon. Since then, the term has appeared in dozens of journal articles, in publications of the Centers for Disease Control, and so on. Sarcopenia is on its way to becoming a first-class disease. As a disease, it will be medically treated. Fifty-year-olds will have access to the treatment, and eventually it will spread to forty-, thirty-, and twenty-year-olds as well.

Cosmetic Gene Therapy

For most of us, though, the real appeal of muscle-boosting gene therapy may not be athletic, but cosmetic. Sweeney, himself a fit-looking thirty-something, sees this as a possible application: "Just say you'd like your pectoralis muscles to be a little larger because you want to look a little better at the beach. Just take a few injections of the virus, and a month later while you're watching television, your muscles have gotten bigger."

In fact, cosmetic uses may very well be more popular than

sports enhancements. In the United States alone, more than 8 million people had some sort of cosmetic plastic surgery in 2001. That's one out of every thirty people or so. In general, the easier a procedure is, the larger the population who'll try it out. Muscle-boosting gene therapy may require just a single injection of gene-toting virus. The most comparable cosmetic procedure in the United States today, the injection of Botox into facial muscles to smooth out wrinkles, is also the most popular. In 2002, doctors in the United States gave patients more than 1.6 million such injections.

The cosmetic applications of genetic technology go beyond bulking up muscles. More than a dozen studies have shown that gene therapy can correct obesity in mice and rats. In 1996, researchers at the Howard Hughes Medical Institute in Chevy Chase, Maryland, gave mice extra copies of the gene responsible for producing the hormone leptin. Leptin regulates body weight by controlling metabolism. Mice given a single gene-therapy injection lost weight while eating just as much as undosed mice. Later studies produced similar results, and also demonstrated that the gene therapy protected the mice against diabetes.

Diabetes and obesity are serious medical issues. Several research groups are working toward the first trials of leptin gene therapy in humans. Yet they also realize that antiobesity gene therapy will have more than medical applications. As Sergei Zolotukhin, a professor at the University of Florida, noted, "This would be the couch potato's dream: You can eat what you want but stay lean." Given the billions of dollars consumers spend on diet products every year, it's almost a given that a onetime injection that kept us thin while keeping us free to eat what we want would be hugely popular.

Another medical therapy with obvious cosmetic applications is the ability to change skin tone. Pale skin is a risk factor for skin cancer. An Australian company called EpiTan is conducting clinical trials of a drug that would give people a tan, and thus resistance to skin cancer, without the need for sunbathing. Gene therapy could darken skin permanently or semipermanently.

In 2001, Heidi Scrable at the University of Virginia created albino mice whose skin can change color. Scrable and her colleagues gave the mice an extra copy of the gene that controls melanin production. However, they used a promoter that would activate only in the presence of specially modified lactose. When the researchers fed the mice the modified lactose, their coats went from white to brown and their eyes from pink to black. When the dietary supplements were withdrawn, the mice went back to albino.

Thus, with the promoter, Scrable and her team not only altered the color of the mice's skin genetically, but also created a system for controlling those genes from the outside. In principle the same thing is possible in humans. You could receive an injection that gave you a tan. Or you could receive new genes that would allow you to change your skin tone, or other traits, via a pill. And it may just be medically viable as a long-lasting therapy against skin cancer.

Researchers are also hard at work on gene therapy as a technique to cure baldness. Robert Hoffman and colleagues at the San Diego biotech company AntiCancer Inc. have shown that they can deliver new genes to the hair follicles of both mice and humans. The researchers inserted genes for green fluorescent protein into skin samples. The hair that grew out of the follicles glowed green under ultraviolet light, proving that it came from the new genes, and thus that it's possible to inject new genes into hair follicles.

Hoffman and his group are now searching for the genes responsible for baldness and gray hair. Once those genes are found, they plan to attempt to alter them in follicles, to see if they can help people retain a full head of healthy hair.

Hoffman's research also points out the more eccentric cosmetic possibilities. There's no reason, for example, that gene therapy couldn't be used to deliver green fluorescence genes to human skin or hair. Such gene therapy would produce humans who glowed under black light. If done using a noninsertional vector, the effects would wear off in a few weeks or months, not much longer than a hair dye lasts.

In fact, other animals have a much wider variety of dyes in their bodies than we humans. Tropical birds and fish are often adorned with bright reds, blues, and yellows, colors that are the result of protein dyes created by the creatures' genes. In principle it would be possible to insert those genes into humans, producing brightly hued skin or hair.

Would society accept these sorts of uses of gene therapy? That's hard to predict. The ultimate limit on what people will do with gene therapy may be simply human desire and imagination. If someone wants to make his skin glow for a few weeks, the technology will be there to allow it. In the meantime, it's likely that more people will be interested in more mainstream cosmetic gene therapies, like permanent increases in muscle mass or permanent changes to hair, eye, or skin color.

Yet the basic techniques of gene therapy are fairly simple. They require only a few thousand dollars' worth of equipment and training that is available at any university. The starter materials—viruses that can be used to convey genes into the body—can be ordered from dozens of scientific supply companies. Thousands of labs in the United States have the expertise and facilities to use them. Any number of graduate students or talented amateurs have the means to set up basement labs in their homes and sell exotic changes to those who want them.

Risks and Benefits

While gene therapy has tremendous potential, it also has risks. Over the last few years, the field has faced two serious setbacks. In September 1999, a young patient named Jesse Gelsinger died, apparently as a result of an experimental gene therapy he'd received for a rare disease called OTC (ornithine transcarbamylase deficiency). Then, in 2002, a French gene-therapy research team led by Alain Fischer and Marina Cavezzena-Calvo discovered that two of their patients, whom they'd just cured of a form of bubble boy disease, had contracted leukemia, apparently as a result of the gene therapy.

In both cases, the problem was the virus used to ferry the genes into the cells of the patients. Viral vectors have two major problems. The first is their effect on the immune system: just as viruses have evolved to penetrate human cells and ferry genes into our bodies, our bodies have evolved immune systems to fight the viruses off. And when those immune systems are triggered, the side effects can be unpleasant, dangerous, and, in rare cases, deadly.

This appears to have been the cause of Jesse Gelsinger's death. Children with OTC lack an enzyme that helps break down the proteins in their diet. Without this enzyme, parts of the proteins combine with hydrogen and form ammonia. The ammonia spreads through the bloodstream and into the brain. In most children it's fatal: often within a month of birth, and usually by age five.

Jesse, eighteen, had a very mild case of OTC—some of his cells could still produce the enzyme. He probably didn't need the therapy himself, but volunteered for it in the hopes that the research would help save the lives of other children born with the disease.

Researchers inserted the functional OTC gene in an adenovirus—a type of virus that causes fever and sore throat. The adenovirus had been stripped of the viral genes that cause the cell to build more virus copies. It had been used on thousands of patients before. It was considered safe. But this time something went wrong.

Perhaps because Jesse was already sick with a different virus, his immune system attacked the adenovirus vector. In the battle between his immune system and the several billion viruses injected into his body, red blood cells burst. The proteins in his blood cells were released into the rest of his body. Normally enzymes in the body would break down these proteins, but Jesse suffered from OTC—he lacked those enzymes. The gene therapy was supposed to fix that. Instead, the unprocessed proteins released by his burst red-blood cells formed huge amounts of ammonia. His blood stopped clotting. His lungs failed. Ultimately his brain stopped working, and his family and doctors decided to turn off his life support.

The second problem, the one that became a factor in the French study, is a lack of precision. Viruses insert the genes they carry into whatever spot in the host cell's genome they can reach. Usually this works out fine. The new gene is inserted in the middle of a fairly harmless or unimportant stretch of the genome. In some cases, however, the newly inserted gene can land in the middle of some important part of the genome—for example, it might end up next to or on top of a gene responsible for fighting tumors.

Think of the genome as a book. The genes are like paragraphs in the book that contain instructions for the cell, telling it what proteins to make and how and when to do so. The gene added by gene therapy is a new paragraph, also containing instructions. Most of the pages of the book are so-called junk DNA, which doesn't seem to serve any purpose. If you insert a new paragraph at some random point, it will probably appear in one of these junk stretches. But occasionally it will appear in the middle of a page with important instructions, or even in the middle of another paragraph. That's when problems can arise.

This is what appears to have happened in the French study. The virus delivered its genetic payload to a location on chromosome 11 known to be important in suppressing cancer. The new genetic material disrupted the function of the existing genes, allowing leukemia to develop unchecked.

Fortunately there appear to be ways around both the immune response and the imprecision of today's vectors. One of the most popular new techniques is to use adeno-associated virus, or AAV, as the vector. Around 80 percent of the world's population is already infected with AAV, yet it appears to cause no disease. Perhaps because of this, the body's immune system hardly responds to AAV, making it a safer vector. It also tends to deposit its genetic payload at a well-known location, away from important cancer-preventing genes, further increasing its safety.

Researchers are also working on a host of nonviral techniques to deliver genes to cells. The simplest is the "naked DNA" technique: just inject copies of the gene into the bloodstream, and some cells will take those genes up. Unfortunately this technique isn't very efficient. Only a tiny proportion of the injected DNA makes it into cells. To improve on the method, researchers have packaged the DNA inside fatty molecules called lipids. They've learned to pack DNA into tiny "nanoballs" small enough to fit through the pores on the surface of cells. They've used a process called electroporation to electrically widen pores in the surface of targeted cells. They've used tiny bubbles to deliver the DNA deep into the body and targeted ultrasound waves to burst the bubbles in precisely the right location to target specific organs. All told, there are at least a dozen such techniques for delivering genes into the body without using traditional immune system–triggering viruses.

In addition, scientists are working on ways to target genetic changes at certain parts of the genome. Going back to the genome-as-book analogy, imagine being able to replace single words or letters at precise locations in the existing book, rather than adding new paragraphs on random pages. A number of researchers have made progress doing just that.

In 2002, a group at the University of Washington led by Roli Hirata and David Russell used an adeno-associated virus to ferry a piece of DNA into the cell. The inserted DNA was almost identical to an existing gene, and the cell's mechanisms detected this and replaced the current DNA sequence with the new, altered one the researchers had inserted. Instead of tacking on a new gene at a random location, Hirata and Russell had managed to replace a precisely targeted genetic sequence with a new one of their choosing. The team then were able to repeat the experiment with a second gene, suggesting that the technique could be used on a wide variety of DNA sequences in the human genome.

Geneticist Alan Lambowitz, at the University of Texas, and his team succeeded with another approach that takes advantage of the "homing" properties of bits of DNA called introns. Introns are a type of so-called junk DNA. Unlike other junk DNA, though, introns actually reside within our genes. Every gene is a long sequence of DNA. Within these DNA sequences are shorter stretches of exons and introns. When the cellular machinery that synthesizes proteins reads your DNA, it uses the exons to decide what kind of protein to build and apparently throws away the introns. Introns almost certainly play some role in the cell, but the exact nature of that role isn't yet understood.

What researchers have discovered is that many introns can home in on specific locations in the genome and insert themselves. If the part of the intron's DNA sequence lines up with DNA already in the cell, it can trade places with the gene that's already there and carry a bit of extra genetic information along with it.

Lambowitz and his team used this characteristic of introns to target them at specific genes. They demonstrated that in human cells they could direct the introns to a chosen location. In bacteria they've used this ability to disrupt dangerous genes. Lambowitz believes they'll also be able to use introns to ferry new genes to a location of their choice. "People who do gene therapy use retroviruses as vectors, which are inserted randomly so there is no control over where they are going. It's a shotgun approach. The intron is more like a rifle shot. You can make it go to the site of your choosing. You can knock out a gene that causes cancer or introduce a tumor-suppressor gene or target a safe site where you will not impair the function of anything else," he explains.

All of these techniques are experimental. There's no guarantee that any one of them will work. The science of gene therapy is still young—less than twenty years old. Its promise lies in what it's already shown us: that it's possible to alter human genes, and that by doing so we can cure disease.

FOR ALL OF the promise of gene therapy to cure disease and sculpt our bodies, there are also several common concerns. The most frequently heard are that altering human genes is unnatural or that gene therapy is unethical because it's not yet safe.

The first argument—that gene therapy is unnatural—rings hollow. It may not be natural to insert new genes into the human body, but the same argument could be—and has been—leveled at almost any medical technique. In 1798, Edward Jenner's vaccine against smallpox was denounced as unnatural and immoral in newspaper editorials and from church pulpits. According to historian Andrew Dixon White, antivaccination societies condemned vaccination as "bidding defiance to Heaven itself, even to the will of God," and declared that "the law of God prohibits the practice."

The use of anesthetics during childbirth was considered "unnatural" and was labeled "meddling in a natural process" until Queen Victoria championed it in the mid-1800s. It's not natural to take blood out of one person and give it to another, yet transfusions are a well-accepted part of medicine. It's not natural to transplant organs, but we've done that with hearts, livers, kidneys, and more. The "natural" thing to do would have been to let Ashanti DeSilva die. Thirty years ago, she would have died within the first few years of her life, simply because we lacked the knowledge to keep her alive. In fact, PEG-ADA, the drug that kept her alive until she could be treated with gene therapy, is also an unnatural product, a synthetic version of an enzyme normally created in the body.

New and unfamiliar medical techniques are often seen as unnatural. Yet with time and usage come familiarity and thus acceptance. By the early 1800s, the smallpox vaccine was in widespread use around the globe. Within years of Queen Victoria's use of chloroform to ease the pain of childbirth, women throughout England were doing the same. Gene therapy may seem threatening today, but as it becomes more effective, as it's used to treat more and more diseases, it will seem less alien and more like any other medical technique we're accustomed to.

The safety argument deserves more consideration. As the cases we discussed a few pages ago make clear, gene therapy still comes with substantial risks. The same is true of every technology we'll discuss in this book. Any attempt to tinker with the human mind or body in a new way is going to have possible complications. Sometimes, as in the case of Jesse Gelsinger, those complications are fatal.

Society could eliminate this risk by simply banning experimental medical techniques. Yet that would end medical progress. Every medical technique has its first human subject. Long before this, scientists and doctors run thousands of trials with animals or cells in a dish to try to ensure safety. Many techniques are rejected at this stage. Only one tenth of 1 percent of new drugs perform well enough in lab and animal experiments to be allowed to reach human trials. Eventually, though, if a medical technique that looks safe and effective in the lab is to benefit real people, it needs to be tried on a first set of human patients.

To be sure, these first human subjects are taking risks—yet that risk is relative to their current circumstances. The first people to go through any particular gene therapy will be those who are extremely ill. They'll be using the technology not to increase their abilities to superhuman norms, but simply in the hope that the experimental therapy will let them live, or to regain abilities that disease or injury have taken away from them. Given the choice between a debilitating disease one knows and a possible cure with unknown side effects, many sick and injured will choose the latter. That's the position Ashanti DeSilva's parents were put in: Watch their daughter's almost certain decline and death from ADA deficiency, or put their hope in an untested therapy? In the DeSilva's case, the gamble succeeded. Ashanti is alive today because her parents were willing to take the sliver of hope offered by a new technique.

Sometimes the gamble fails. The first heart transplant patient lived just 18 days after the operation. Barney Clark, the first man to receive an artificial heart, died 112 days after the implantation of the device. Many of the first humans who underwent other groundbreaking medical techniques died either as a result of complications or simply because the therapy failed. Virtually every new medical technique has problems that need to be solved by further trials. Yet the first human trials are not in vain; they're an indispensable part of medical progress.

The experimental gene therapy that killed Jesse Gelsinger taught researchers about the risks of immune response to viral gene vectors and spurred research into other, safer techniques. Gelsinger won't benefit from those lessons, but the next generation of children born with OTC might. Jesse understood his role. Before his therapy he told a friend, "What's the worst that can happen to me? I die, and it's for the babies."

Again and again, we see that with new medical technologies, the early adopters are those who have the most to gain and the least to lose. In some cases, those first patients benefited tremendously—their life prospects have been transformed. In others, they've suffered side effects. Yet in every case, they've provided scientists and doctors with the opportunity to learn more about the human mind and body, leading to better, safer techniques that can heal millions of other sick and injured.

Ironically, one of the most obvious responses to safety concerns about enhancements, banning enhancement techniques, is counterproductive. Enhancement techniques are likely to be quite popular. Consider some of the precedents: In addition to the more than 8 million cosmetic plastic surgery cases they underwent, U.S. consumers in 2002 spent $17 billion on sports supplements and herbal products intended to improve health or boost mental or physical function. Most of these supplements have little or no effect, yet they're incredibly popular. When truly effective mental and physical enhancements are available, they'll tap into a large existing demand.

Bans on highly in-demand goods and services don't seem to eliminate the market for such things—they simply move it underground. That's what happened in the 1920s when Prohibition in the United States made alcohol illegal, and it is also the case today with the War on Drugs. Neither effort stopped the flow of the banned substance; instead they led to the creation of black markets for the products people sought.

In a black market, safety suffers. There are no regulators to enforce safety standards. There's no threat of liability for botched services or procedures. It's difficult to perform long-term safety studies to spot emerging problems.

For example, illegal drugs sold on the street are often not what they're claimed to be. In 1999, drug safety researchers determined that the majority of emergency room visits in Oakland, California, that were attributed to use of the drug ecstasy (MDMA) were actually the result of dangerously high doses of dextromethorphan (DXM), an ingredient found in cough syrup, which had been pressed into tablets and falsely sold as ecstasy. In a regulated market, this kind of adulteration and misrepresentation seldom occurs. Inspectors and safety checks largely guarantee that the consumer gets what he or she is paying for.

Another example is the effect that bans on abortion have on safety. In countries where abortion is legal, only one out of every hundred thousand abortions leads to death. In countries where abortion is illegal, the death rate can be seven hundred times higher. The World Health Organization estimates that worldwide, one out of every eight maternal deaths is due to an unsafe abortion, most of them in areas where abortions are illegal. When a service is made illegal, many of the safety standards regulators uphold go by the wayside. If safety of enhancement techniques is a concern, it makes more sense to regulate them than to ban them.

That regulation must also take into account the use of these techniques as enhancements, not only as medical procedures. The cyclists who died from overdoses of EPO died because they lacked reliable information on how much was safe. The literature around EPO and the safety tests that EPO products had to go through prior to regulatory approval were focused on the medical uses of the drug. The studies involved animals and patients with anemia, not healthy men and women, let alone athletes at the top of their game. Banning the enhancement use of EPO didn't stop cyclists from using it, it just robbed them of information that might have saved their lives. A truly effective safety system must acknowledge that people will use medical products as enhancements, and establish safety guidelines for those uses.

In just a few decades, we've gone from the first tinkering with human genes to the discovery of dozens of techniques that could alter the human genome in very precise ways. Those techniques give us the power to cure diseases or to enhance and sculpt our bodies. This new control over our genes promises to enhance our quality of life as dramatically as the medical discoveries of the past century.

CHAPTER 2

Choosing Our Minds

NE OUT OF every ten people aged sixty-five or older has Alzheimer's disease. Nearly half of those over age eighty-five have it. Worldwide there are more than 12 million sufferers of the syndrome. By 2025, that number is expected to almost double, to 22 million. In the United States alone, consumers and insurance companies spend over $100 billion on the disease each year, most of it on medical and nursing care.

Alzheimer's devastates individuals and families. As memory and competence fade, Alzheimer's patients become dependent on their spouses, their children, or whoever can be found to care for them. For the caregivers, it's a long slow process of watching a loved one fade away, compounded by the often crushing financial burden of caring for their relative.

Small wonder that so much effort is being put into finding a cure. The National Institute of Aging spends half of its billion-dollar annual budget on research into Alzheimer's, and dozens of private companies and university labs funded by other sources are conducting research on the disease as well.

One of the more promising approaches is gene therapy. On April 5, 2001, Mark Tuszynski and other researchers at the University of California at San Diego (UCSD) implanted genetically modified neurons into the brain of a sixty-year-old woman suffering from Alzheimer's; the neurons had been modified to have extra copies of the gene responsible for production of nerve growth factor, or NGF.

NGF is a chemical that triggers the growth of neurons. As we age, levels of NGF in the brain slowly drop. This may explain why neurons in aging brains tend to shrivel up and die. In particular, the axons of neurons, the long fibers that extend out of them and signal other neurons, tend to shrink.

In a series of earlier studies, Tuszynski had found that adding extra copies of the NGF-producing gene not only prevented this age-related shrinkage in the brains of mice and monkeys but could actually *restore* the neurons in old brains to their youthful size, shape, and activity. Persuaded by Tuszynski's findings in animals, the Food and Drug Administration approved his proposal for human trials. It is one of the few human gene-therapy trials to target the brain.

After following his patients for years, Tuszynski and his team found that the NGF gene therapy, although it didn't cure Alzheimer's, slowed the rate of onset by a factor of three—making it one of the most effective tools against Alzheimer's yet developed. If larger trials find the same results, the therapy could be used on a large scale within the next several years.

Extra NGF may do more than merely restore lost brain function. In 2000, Howard Federoff and colleagues at the University of

Rochester discovered that NGF gene therapy could actually *improve* the learning and memory of normal mice. Federoff compared normal mice to mice genetically engineered to produce more NGF in one part of their brains. In a maze-navigation test, by the third trial mice with extra levels of NGF navigated the maze about 60 percent faster than normal mice.

In another test, the presence of extra NGF copies made an even bigger difference among mice that were raised in a more intellectually challenging environment—one with more toys and obstacles, more puzzles and tests for them to complete. NGF-enhanced mice raised in the stimulus-rich environment learned the maze *four times as fast* as normal mice in the stimulus-rich environment, and more than *five* times as fast as normal mice raised in ordinary cages.

Learning to navigate a maze is one thing. Human learning— with its focus on language, mathematics, abstract ideas, and interpersonal relations—is more complex. Nevertheless, the NGF-enhanced mice illustrate the power of genes to enhance learning and memory, and a number of other studies carried out on different genes have produced animals with just as large a mental edge.

Princeton biologist Joe Tsien's "Doogie" mice are probably the most famous of these "smart rodents." Named after TV's child prodigy Doogie Houser, M.D., the mice were featured in *Time* magazine and on *The David Letterman Show.* On a variety of learning tests, they perform better than normal mice—in some tests as much as five times better.

To understand what Tsien did, we have to understand a bit about how the brain works. Your brain has approximately 100 billion neurons—fifteen times as many neurons as there are people living on this planet—and on average each of those neurons connects to about one thousand other neurons. This adds up to around 100 trillion of those connections between neurons, called synapses.

Synapses are points where messages are conveyed from one neuron to another. Messages generally come in the form of individual molecules called neurotransmitters. One neuron, the presynaptic one, fires. It releases a flood of these molecules. The message is conveyed when these molecules bump into the matching receptors on the postsynaptic neuron, the one on the listening end. Receptors are like three-dimensional puzzle pieces. They're shaped so that only certain neurotransmitter molecules will fit well inside them. So for the signal to be sent, the neurotransmitters released from the presynaptic neuron have to be the right kind to dock snuggly with the receptors on the postsynaptic neuron.

The most common kind of neurotransmitter molecule is glutamate. Neurons that send signals with glutamate are found in just about every nook and cranny of your brain. Tsien and others have focused mostly on glutamate-using neurons in the part of the brain involved in long-term memory, called the hippocampus.

Neurons in the hippocampus, as in most of the rest of your brain, experience something called long-term potentiation, or LTP. When two neurons fire in sync, the connection between them is potentiated, meaning it gets stronger. This potentiation, this strengthening, can last for weeks or months or even years— hence the phrase *long-term potentiation*.

Tsien knew that as humans age, LTP becomes less and less frequent. It takes more signals between neurons to convince them to strengthen their connections. He also knew that as we age, our brains produce less of a protein called NR2B, which is a crucial part of the NMDA receptor—one of the receptors found in the hippocampus that respond to signals from glutamate. Tsien reasoned that these two phenomena could be related, and could explain why young people have better memories than older people. That led him to genetically engineer a strain of mice with extra copies of the NR2B gene—the Doogie mice.

The results surprised even Tsien. The Doogie mice didn't just retain their memories in old age, they actually formed memories— *learned things*—more quickly at any age. After being shown an object for five minutes, Doogie mice remember it five times longer than normal mice. When placed in a water-filled maze, Doogie mice learn twice as quickly as normal mice how to find the hidden platform they can remain dry on. When given an electrical shock in a room, the Doogie mice are about 30 percent more likely to remember it ten days later. If researchers turn off the electrical current in the room, Doogie mice figure that out twice as quickly as normal mice. This last point is important. Not only do the Doogie mice learn more quickly than normal mice, they also *unlearn* more quickly.

Faster learning and better memories aren't necessarily free blessings, though. The NMDA receptor is involved in the response to addictive drugs like cocaine and heroin. NMDA is also involved in stroke. That raises the possibility that people taking NMDA-affecting drugs or gene therapy could be at greater risk for drug addiction or stroke. Later studies also suggested that Tsien's mice feel major pain longer than normal mice. After being injected with a noxious substance in a paw, the Doogie mice lick at it longer than other mice.

Despite these possible problems, interest in applying Tsien's research to humans has been intense. Indeed, in 2003, researchers discovered that Alzheimer's patients had lower-than-normal activity of the NR2B gene, suggesting that Tsien's research could be directly applicable to treating the disease. A number of other companies have been formed with the intent of creating memory-improving drugs based on related parts of the memory system.

Other groups are working on genes that affect what goes on inside the cell after the NMDA receptor is activated. In the 1980s, Nobel Prize–winning neurologist Eric Kandel discovered that a protein called CREB (cyclic amp-response element binding) is involved in the changes inside a neuron that build and strenthen synapses.

In 1994, a competing researcher named Tim Tully built on Kandel's work. Tully and a colleague at Cold Spring Harbor Laboratory on Long Island added a gene to fruit flies to increase the amount of CREB in their brains. The resulting fruit flies had astoundingly good memories. For example, they could learn to associate a smell with an electrical shock in one repetition, while normal flies took ten tries to learn the same thing. Interestingly enough, the genetically engineered flies didn't do any better after more repetitions. After ten repetitions the normal and engineered flies performed the same. But the flies with more CREB got to the same level of performance as the normal flies in one tenth the time.

Experiments in mice and sea slugs have shown the same effect as in fruit flies. Animals genetically engineered to have higher levels of CREB learn things faster. The genetically engineered animals also seem perfectly healthy—aside from their enhanced learning ability, they behave just like other members of their species.

In 1998, Kandel leapt ahead when he showed that a drug that boosted the amount of CREB in the brain could improve long-term memory in mice. Kandel found that injecting mice with Rolipram, an antidepressant that indirectly raises levels of CREB, could boost mouse memories. In tests of their memory twenty-four hours after training, Rolipram-injected mice did about 50 percent better than mice injected with a saline solution. Tully did his own experiments and found that mice injected with Rolipram could learn things in two trials that other mice required five trials to get right.

Emboldened by success, Tully and Kandel went on to found competing pharmaceutical companies. Rolipram, as it turns out, is not a good drug option for most people. Its side effects include nausea and vomiting. Kandel's company, Memory Pharmaceuticals, and Tully's company, Helicon Therapeutics, are both working on new drugs that work similarly to Rolipram but which can be taken as pills instead of injections, and don't have the nasty side effects. Both drugs, MEM-1414 and HT-0712, are just entering human trials.

In total, at least a dozen companies in the United States alone are pursuing some sort of memory-enhancing drug today. The motivation is a large and glittering market. Four and a half million of the world's Alzheimer's patients live in the United States. In addition, there are another 4 million or so Americans with mild cognitive impairment, or MCI, a syndrome the Food and Drug Administration now recognizes as a treatable disease. Every year, around 10 to 15 percent of MCI patients degenerate to the point of manifesting full-blown Alzheimer's disease. Most recently, the FDA has recognized "age-associated memory impairment," or AAMI, as a treatable disease. Anyone over age fifty who is in the bottom 16 percent of his or her age group in memory tests is defined as having AAMI—that's 12 million people in the United States alone.

The market may be larger yet. Tully notes that there are now almost 80 million people aged fifty or older in the United States and that this age group will double by 2030. And UC Irvine's James McGaugh, a critic of memory-enhancing drugs, told *Forbes* magazine that "drug companies won't tell you this, but they are really running for the market of nonimpaired people—the forty-four-year-old salesman trying to remember the names of his customers." Axel Unterbeck, the president and chief science officer of Memory Pharmaceuticals, concurs, saying, "If [the treatment] is safe, the market is incalculable."

Ultimately, individual people will judge the pros and cons of memory enhancers. That process is starting already, with the first few human trials of the current round of drugs. If those trials are successful, more studies will follow. Eventually, elderly patients may start using the drugs. If they report satisfaction, usage will spread to a wider population. If the benefits of the drugs outweigh their cost and side effects, they may be used in the way Tully suggests: "Memory enhancers could become 'lifestyle' drugs to be used by anyone interested in learning a second language, in playing a musical instrument, or in studying for an exam."

Of course, the current crop of drugs could fail in human trials. The most likely stumbling block would be unpleasant side effects. If, like Rolipram, they make their users physically ill, they're unlikely to become popular. But whether or not the current batch of drugs succeeds in the market, the last decade has taught us a tremendous amount about the molecular basis of learning and memory. Sooner or later, someone will figure out how to make a drug that improves human learning without serious side effects. As Tully told *Forbes,* "It's not an 'if'—it's a 'when.' "

Smart Drugs—Here Already

Smart drugs and smart genes may seem far-fetched, but in fact they're already so entrenched in our culture that we hardly notice them. Hundreds of millions of people already use some sort of drug to improve their memory or attention. Caffeine and nicotine are so familiar to us that we don't think of them as performance enhancers, but that's exactly what they are. Every time you have a coffee or caffeinated soft drink to stay awake while driving, you're engaging in a mental enhancement: pushing your alertness beyond its normal human endurance.

Caffeine and nicotine slightly improve reaction time and attention. Nicotine also improves long-term memory and performance on problem-solving tests. As any smoker will tell you, nicotine helps people concentrate. Researchers at Stanford University have even found that well-rested pilots, when given nicotine before a flight simulation test, perform better than control subjects. The biggest improvement is in complex, attention-demanding tasks like landing the plane and responding to simulated emergencies.

Currently nicotine usage is dropping dramatically, but the socially sanctioned use of other, more powerful cognition enhancers is on the rise. In 1999, physicians in the United States wrote more than 15 million prescriptions for the stimulant drugs Adderall (amphetamine), Ritalin (methylphenidate, a close cousin of amphetamine), and their clones. Almost all of those medications were prescribed for attention deficit hyperactivity disorder, or ADHD.

Studies suggest that ADHD is diagnosed and drugs for it prescribed too often. Yet it's clear that most children and adults who are diagnosed with the disease respond well to the stimulant drugs that are prescribed for it. Generally, Ritalin and other drugs improve attention, memory, test scores, and grades. For children with the most severe ADHD symptoms, these drugs reduce the risk of legal trouble, of dropping out of school, and even of later substance abuse.

What's less appreciated is that drugs like Adderall and Ritalin improve attention and memory in almost everyone who takes them, whether they are diagnosed with ADHD or not. NIH researcher Judith Rapoport has found that ADHD drugs give the same mental boost to normal children and adults as they do to hyperactive ones. As psychologist and professor Ken Livingston has written in *The Public Interest,* "Even if you have never been diagnosed as having a problem paying attention, many of these drugs will improve your focus and performance." Lawrence Diller, author of *Running on Ritalin,* agrees, calling Ritalin a "universal performance enhancer" and noting that when normal children are placed on the drug, they develop better-than-normal memory and attention.

Another drug that's recently caught the public eye is modafinil. Better known by its trade name, Provigil, modafinil is a French drug approved for use in treating narcolepsy. Narcoleptics are prone to sudden and intense drowsiness and even uncontrollable bouts of sleep. Modafinil gives narcoleptics the ability to remain alert and awake throughout the day. It can also be used to keep you awake for extraordinary lengths of time—up to several days in a row. While awake, modafinil users suffer fewer jitters and other side effects than people given stimulants like caffeine or amphetamines.

People under the influence of modafinil, unlike stimulants, can sleep when they choose to. When they do eventually sleep, they suffer less of a "rebound" than people kept awake for the same amount of time on amphetamine or nothing at all. For these reasons, the U.S. military is testing its use for pilots and other soldiers as part of its Continuous Assisted Performance program.

The other ubiquitous mental enhancer of our time is the antidepressant. Almost 38 million people in the world today are prescribed Prozac, and at least as many are taking some closely related drug. We don't usually think of antidepressants as enhancers. After all, their real goal is to restore someone to a more "normal" state of mind. The reality is that Prozac and its clones affect both the depressed and the nondepressed.

For example, a 1998 study by Victor Reus and Brian Knutson at the University of California at San Francisco found that that the antidepressant Paxil made already perfectly happy people more socially adept in challenging group situations. In group problem-solving studies, people given Paxil were rated as more effective partners by judges who didn't know who'd been given the drug and who hadn't. Another study that year showed that Paxil reduces anger in nondepressed individuals. Animal studies have shown that monkeys given drugs like Prozac are more likely to rise to the top of their social hierarchies in times of turmoil.

Antidepressants and ADHD drugs have their downsides. Prozac interferes with orgasms. Ritalin and Adderall can produce jitteriness, headaches, irritability, and increased blood pressure. In extreme cases, they can turn a formerly creative, adventurous child into a listless or depressed one.

Most seriously, Ritalin, Adderall, and their ilk can easily be abused, even to the point of addiction. Adderall, the second most prescribed drug for ADHD, is actually just a trade name for pure amphetamine, and Ritalin is closely related. Both have the same general effect on the brain as methamphetamine ("crystal meth") or cocaine. All four drugs mimic the effects of the neurotransmitter dopamine, which your brain normally releases in response to extreme pleasure or to reinforce positive behaviors. When you take a dopamine-mimicking drug, the brain reinforces the action of taking the drug. Frequent use at high doses increases the reinforcement.

Eventually, gene therapy may provide a long-lasting or permanent alternative to these powerful drugs. Researchers have already linked a mutation of the gene that regulates one kind of dopamine receptor to a heightened risk for ADHD. They've identified two genes on chromosome 13, G30 and G72, that increase the risk of bipolar disorder, or manic depression. Researchers have also known for years that mutations in a gene responsible for the serotonin transporter, the part of the neuron that sucks serotonin back in and recycles it, are associated with depression, anxiety, and binge drinking.

At some point, doctors may use gene therapy to treat depression, ADHD, or bipolar disorder. So far that hasn't been tried, but gene-therapy experiments in both animals and humans have shown some interesting results in other cognitive areas.

One of these is pain management. Chronic pain is an immense problem for cancer sufferers, amputees, and burn victims. The drugs that are most effective in controlling that pain, opiates like morphine, have serious side effects. They dehydrate patients, make them drowsy and constipated, and can lead to physical addiction. A team at the University of Pittsburgh led by David Fink is trying to find ways around that. In 2005, Fink's team is set to launch the first human trials of gene therapy to control pain. In previous studies, Fink and his colleagues gave mice extra copies of the gene for proenkaphalin, a naturally occurring painkilling molecule. The mice whose bodies produced more proenkaphalin showed a higher pain threshold than genetically normal mice, and they didn't manifest any of the side effects of opiate drugs. Numerous other groups have seen similar results. Fink hopes that the research will lead to techniques to control chronic pain in patients.

Another intriguing area for gene-therapy research is the treatment of alcoholism. Researchers have known for years that low levels of a certain type of receptor for the neurotransmitter dopamine (the D2 receptor) are associated with increased risk of alcohol addiction. In 2001, Panayotis Thanos and Nora Volkow at Brookhaven National Labs tried to see whether they could correct that condition through gene therapy. The researchers delivered extra copies of the gene for the D2 receptor into the brains of rats that had normal levels of D2, but that had become addicted to alcohol anyway. The neurons in the rat brains grew more of the dopamine receptor, and in the days that followed, the most profoundly addicted rats cut their drinking habits by two thirds.

Gene therapy might even be used to affect our personal relationships. In 2001, Larry Young and colleagues at Emory University in Atlanta showed that new genes could change mating behavior. The researchers gave male prairie voles gene therapy to increase the number of receptors in their brains for the hormone vasopressin. The voles that received gene therapy bonded to female voles more quickly than normal, even without the usual prerequisite of sex. They were also more friendly with other males and less likely to take risks.

Researchers already know that vasopressin is involved in

human romantic bonding. With oxytocin, it's one of the brain chemicals whose levels increase over the course of a romance. Something as simple as a kiss can release vasopressin into your bloodstream. The finding that changing the genes for vasopressin can change the mating behavior of other mammals dovetails nicely with this, and suggests that men and women could chemically alter their relationships if they so chose, either temporarily, with drugs, or permanently, with gene therapy. Someday clinics might offer couples gene therapy to strengthen and extend their emotional bond.

Many other genes are associated with various aspects of human behavior—genes that we might be able to use to sculpt our personalities. For example, variations in the gene for a brain chemical called monoamine oxidase A (MAOA) are associated with thrill-seeking and novelty-seeking behavior. We know that the 5HT2-C serotonin receptor is involved in a variety of spiritual experiences. There's the still-controversial claim that a gene somewhere in the Xq28 region of the X chromosome increases the likelihood of homosexuality. At the other end of the spectrum, researchers Karen Carver and Richard Udry at the University of North Carolina have calculated that genes account for 26 percent of the variation in strength of religious belief.

In the next few decades, this accumulated knowledge base could be used to create new drugs that sculpt or alter any aspect of human behavior: infatuation, pair bonding, empathy, appetite, spirituality, thrill seeking, arousal, even sexual orientation. If and when gene therapy in the brain is ever feasible, it will be possible to opt for permanent or semipermanent alterations of personality. As a species, we're in the process of figuring out what makes us tick. The next step—a step tens of millions of consumers will pay for—is to turn that knowledge into products that allow us to sculpt our own minds.

Neurotechnology and Society

Greater power over the human brain and mind will have profound effects on society. New mind-altering drugs and gene therapies hold promise for treating Alzheimer's disease, age-related decline, chronic pain, and many other conditions. Smart drugs and smart genes alone may improve the quality of life of tens of millions of people who currently struggle with mental illness.

The benefit in social and economic terms may be just as large. Americans spend almost $100 billion per year on long-term care for Alzheimer's disease patients. According to the National Institutes of Mental Health, other mental illnesses, including depression, cost the United States almost $300 billion per year in lost worker productivity, medical care, secondary illnesses, and so on. Reducing the incidence of Alzheimer's, depression, and other major mental diseases would save America tens of billions of dollars annually. Actually improving the learning speed and attention of mentally healthy people may yield productivity gains of hundreds of billions of dollars more.

People with better memories and quicker minds will earn more money and produce more for others. Any technique that increases the human ability to learn, to think, or to communicate is going to produce economic returns. It will increase our ability to solve problems, to make scientific breakthroughs, and to build better products. Scientists who learn more quickly will be better able to stay abreast of developments in their fields. Doctors and nurses who can stay alert longer will make fewer errors in treating patients. Smarter engineers will produce better products to improve our lives. Smarter programmers will make better software. Smarter architects will design better buildings. Smarter biologists will come up with new medicines. Overall, society will become richer.

This isn't just speculation—it's supported by years of study. Around the world, the correlation between the average IQ of a country and its GDP is an incredibly high 0.76. The correlation between average IQ and rate of economic growth over the last quarter of the twentieth century was 0.64. According to Robert Barro of Harvard, a single standard deviation gain in science and math test scores raises economic growth rates by around 1 percent. In the United States, that's a gain of $1 billion per year to the economy for every point gained on the SAT.

Another study, by the British Association of University Teachers, found that every additional year of education across the population adds around half a percentage point to annual economic growth rates; and that the percentage of a country's GDP that it spends on education has around a 0.5 correlation with its wealth. Stronger memories that reduce the amount of time needed to learn a subject will make that educational investment more effective. Smart drug– or smart gene–enhanced students will be able to learn more and spend less time in school.

That improved economic growth isn't an abstract—it's real money enabling real improvements in the quality of our lives: better homes, safer cars, more effective medicines. Improving the human mind improves all the products of the human mind. Any technique that improves on the human intellect will pay dividends throughout society.

This has competitive consequences as well. It's possible that in the near future, some countries will ban mind-enhancing drugs and other mental enhancement while other countries will embrace them. For example, the United States and Western Europe might prohibit any nontherapeutic mental enhancement while in China, India, and much of Asia they become commonplace.

Asian countries are in general more friendly to the idea of human enhancement than Western countries. For instance, international polls of attitudes toward genetic engineering find that while only about 20 percent of Americans approve of the use of genetic techniques to create desired traits in children, 63 percent of the population of India and a whopping 83 percent of those polled in Thailand approve. That poll was specifically about genetic engineering, but the numbers suggest some fundamental attitudes toward enhancement. If drugs or gene therapies that boost mental performance are outlawed in the United States, they may still move forward in other countries, and the nations that embrace them will gain a competitive advantage over those that do not.

With the benefits of enhancement come some potential drawbacks. Deeper knowledge of the brain will make it easier to develop new recreational drugs. Many of those will be quite harmless, even enriching. Others may be addictive and dangerous if abused.

The power to control our emotional states also raises questions about our sense of identity. Once the technology is mature, some people will choose to alter their personalities in ways that their friends and loved ones don't understand. Some people may seek out ways to become more aggressive, or violent, or hostile, or to suppress their capacity for guilt, remorse, and empathy.

Society isn't likely to sanction the more radical of these changes, yet as the technology matures, it will be easier and easier to obtain it on the black market. There's no reason why, in the long run, splicing a gene into a new vector will be any harder than cooking up a batch of illegal drugs. And as we can see, there are plenty of chemists with the skill and motivation to do the latter. Where there is demand for an illicit mental alteration, there will be supply.

Even the ability to alter our minds less drastically will bring us face to face with some deep and perplexing questions about who we are. Am I still the same person if an injection changes parts of my personality—if I go from shy and reserved to being a risk taker and thrill seeker? What if I always wanted those traits for myself? What if my friends no longer like me, or if I no longer like them? If my wife and I take injections to increase the intensity of our neurochemical bond, what does that say about our relationship? Has our love been deepened or demeaned?

These questions can be asked without any new technology. Am I the same person under the influence of alcohol or other drugs? Am I the same person I was ten years ago? If I learn a new language, move overseas, make new friends, get a new job, or change my religion, am I still the same person I was before?

Neurotechnology doesn't radically alter the nature of identity—it just brings some of the limitations of the idea into starker relief. The reality is that we're constantly changing. Every experience we have alters us—intellectually, emotionally, neurobiologically. As you read this book, electrochemical impulses are racing between the neurons in your brain. In those neurons, genes are being turned on and off. Dendrites are stretching out to close the gap between neurons firing in sync. Other connections are weakening and separating. Your brain will never be the same as it was before you read this page.

At the psychological level as well, permanent changes are being made. Memories are being formed. Ideas are being triggered.

Older thoughts are resurfacing. Your future behavior is being altered slightly. What you've read and your reactions to it may affect your choices in other books or movies or conversations tomorrow or next week or next year.

These changes are all happening because of a choice you made. You chose to pick up this book. The other choices you make— what to read, what to watch, where to go, whom to spend time with—affect your mind and brain as well. Some choices—what to study in school, where to live, what job to take—can have a profound impact for years to come. To a large extent, the choices you make today affect who you'll be in the future.

Neurotechnology will present us with choices that differ only by degree from all of life's other choices. It will make this power to choose who we are more obvious and immediate. That in turn will make individual responsibility for our behaviors and identities more clear. In a world where we can sculpt our own emotions and personalities, people will no longer be able to say "I can't help it, that's just the way I am." Undoubtedly the transition to that level of individual choice will be difficult, yet in the end it seems to be a fundamentally positive development—an increase in our power to determine who we are and to take greater responsibility for who we become.

CHAPTER 3

Created Equal

N HIS 1997 commencement address at Morgan State University, President Bill Clinton said, "Science and its benefits must be directed toward making life better for all Americans—never just a privileged few. Its opportunities and benefits should be available to all."

Clinton's statement could be stronger. For science to be just, one could argue, its benefits must be available to everyone around the world, not just those in rich Western countries.

The power to sculpt our minds and bodies is an outgrowth of our efforts to heal. The same is true of technologies that can slow human aging, alter the genes of our children, and link the abilities of computers with our brains. All of these enhancements emerge from medical research, so we can turn to medicine for an assessment of the likely economics of human enhancement.

The history of medical technology and its prices suggests two reasons to believe that enhancement technologies could become cheap enough for most of the world to afford. First, although most enhancement techniques we'll discuss are expensive to develop and test, they're cheap to manufacture, so that after initial research and development, few further investments must be made. Once initial investments have been recouped, prices can fall. Like almost any new technology, human enhancement techniques will at first be available only to the rich, but over time, they'll become more affordable, allowing more and more people to reap their benefits.

Second, spending on biotech enhancements is likely to obey the law of diminishing returns: someone who spends ten times as much on an enhancement is going to get much less than ten times the result. The biggest gains are going to come from relatively small amounts of spending. Everything beyond that will produce only small refinements.

Let's focus first on price. Consider the cost of a prescription drug. When you walk into a pharmacy and fill a prescription, what are you really paying for? To you it may seem that you're paying for the substance contained in those pills or tablets. After all, what you care about is putting that chemical in your body and enjoying the beneficial effects—a reduction in pain, the elimination of some disease, or an alleviation of depression—but in a sense, what you're really paying for is information.

Drugs cost almost nothing to manufacture. Generally, once researchers figure out that a certain molecule causes a certain effect in the body, almost any laboratory can synthesize the substance at a reasonable cost. And the more of the substance it makes at once, the lower that cost goes.

The real cost is in the research and development that leads to a new drug—thousands of experiments, years of work, dozens of other candidates thrown away, and long and costly safety trials to get approval from the Food and Drug Administration to market the drug. According to a 2000 study done at Tufts University, the average cost of developing and approving a new pharmaceutical is now close to a billion dollars.

Pharmaceutical companies don't make this investment out of altruism. They're motivated by the potential profits, and those profits are protected by drug patents. As long as the patents on a new drug are in effect, only the company that holds the patents can determine who is allowed to manufacture the drug. That company can charge as much as it thinks customers are willing to pay.

Drug patents last for twenty years after the invention of a drug. Often it takes a decade or more for a drug to reach the market after its invention, so for practical purposes a patent preserves a monopoly for the manufacturer for a decade or less. Once the patents expire, other companies can manufacture and sell generic versions of the drug, without having to spend the huge amounts to develop it again from scratch. When multiple companies sell the same drug, they're forced to compete on the basis of price, which drives the cost down for consumers. The greater the demand, the lower the price goes. In addition, higher demand leads to greater economies of scale in production and increased price competition as more companies manufacture the drug. According to a 1998 Congressional Budget Office study, the cost of drugs whose patents have expired and that are manufactured by more than twenty companies is on average one fifth that of drugs that are still patent-protected. We can see this in Figure 1 (page 64).

In a way, what's happening here is a shift in control from corporations to consumers. While patent protection lasts, the inventor has complete control over how much of a drug is made and how

Average Pharmaceutical Cost by Number of Sources, in 2000

FIGURE 1 As more sources manufacture a drug to meet consumer demand, prices drop.

much to charge for it. Once patent protection runs out, manufacturers must scramble to outdo each other in bidding for the customer's dollar.

How low can that competition drive the price of a drug? Again, the greater the demand, the lower the price. Penicillin was one of the greatest breakthrough drugs of the past century. At first, it was extremely expensive. For decades after its discovery and development as an injectable drug in 1941, the drug was effectively priceless, for there was no method to produce it in bulk. With the demand for antibiotics of World War II, large-scale synthesis methods were developed. By the end of the war, the price of penicillin had dropped to around twelve dollars per gram. Over the next thirty-five years, it continued to drop, to a current cost of less than two cents per gram, or less than one five-hundredth its price in 1945.

There's no reason to believe that new enhancement drugs will be substantially more expensive to produce than penicillin if manufactured in equivalent volume. Gene therapies will be a bit more expensive at first, solely because they're manufactured through new and different techniques. But fundamentally they're information products in much the same way that drugs are. The cost of manufacturing is almost insignificant when compared to the cost of research, development, and testing.

What does this tell us? It means that the first effective performance-enhancing drugs may well be expensive at first. A company that had a drug that doubled your learning rate could charge an incredible amount for it. A company that had a drug that cut your rate of aging in half could charge even more.

Yet once the patents on truly effective performance-enhancing drugs have expired, their prices will drop in proportion to the number of people who want the drug. The most sought-after enhancements will be the cheapest. A drug that significantly slows aging, for example, might be in as much demand as penicillin, and therefore just as cheap—pennies a day. If so, such a drug would be within the buying power of most of humanity.

Of course, not all enhancements take the form of drugs or injections of gene vectors. Some enhancement techniques will require considerable time or equipment. Yet over time, these sorts of procedures drop in cost as well. Consider laser eye surgery, specifically the LASIK technique used to correct poor vision. Initially there were only a few LASIK clinics in the world. Those clinics often had no competitors in the same city. In the first year that LASIK was available, only twenty thousand procedures were performed, and the average cost per eye was around $4,000. But as more ophthalmologists entered the business of providing LASIK, the cost dropped and the number of procedures soared. By 2000, more than 1.5 million eyes were being sculpted by laser each year, and the cost was as low as $500 per eye. Today some clinics charge as little as $200.

Two hundred dollars is still a lot of money for most people on earth. Yet the cost of LASIK has dropped as much as it has while the patents that cover it are still held by just two companies: VISX and Nidek. When the first of those patents expires, in 2007, the cost is set to drop even further, bringing it within reach of many more people. This pattern suggests that labor-intensive enhancements are also controlled by a kind of inverted law of supply and demand: Greater demand drives prices up in the short term, but down in the long term. Highly in-demand services will attract providers, and the competition among those providers will drive prices down, shifting control to consumers.

Not all medical procedures follow this pattern. Most surgeries are paid for by insurance. Since insurance companies set the price they're willing to reimburse, and the patients choosing the doctor seldom have much awareness of the overall price, there is not the same competition to push costs down. However, for purely *elective* procedures where the patient pays the price, the rule applies: Greater demand drives prices up in the short term, but down in the long term.

The other reason to believe that legal enhancement technologies could benefit the poor almost as much as the rich is the law of diminishing returns. The better off you already are, the more you have to spend to increase your well-being overall.

Consider automobiles. In most places in the United States, a few thousand dollars will purchase a fairly safe, comfortable vehicle for getting from point A to point B. Alternately, many tens of thousands of dollars will buy you a new luxury vehicle. A luxury car can cost ten times what the most affordable new economy car would cost, yet the luxury car isn't ten times as fast, or ten times as fuel-efficient, or ten times as safe. The extra money spent on it buys only incremental advantages. In terms of basic mobility, the inexpensive step between no car and a cheap car is larger than the very expensive step from a cheap car to a BMW.

This law of diminishing returns applies to other aspects of life as well. One good measure of human well-being is life expectancy. To have a high life expectancy, you must have access to food and water, adequate shelter from the environment, easily available health care, vaccination against major diseases, protection against violence and warfare, an environment free of pollution, crowding, and stress, and so on. What's more, life expectancy correlates well with education, with political freedom, with equality within a country, and with other positive social indicators. Overall it may be the best single proxy for the various components of human quality of life.

Like the quality of a car, life expectancy goes up as you spend more money, but more and more slowly as spending increases. Figure 2 (page 68) shows the life expectancy and per capita gross domestic product (GDP) of 175 countries. Higher per capita GDP moves a country toward the right. Higher life expectancy moves a country toward the top. The trend line that cuts through the dots initially shoots up toward longer life, and then it flattens out. Greater national wealth strongly affects life expectancy at the low end. But above per capita GDP of about $3,000, more money has only a small impact on life expectancy.

The same pattern is generally true if we look at life expectancy vs. per capita health spending instead of per capita GDP. Going

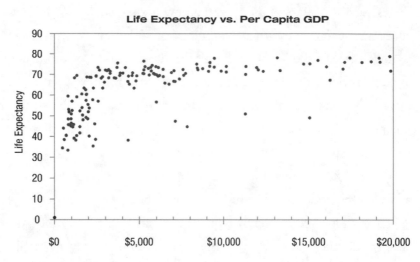

FIGURE 2 Life expectancy rises quickly with increases in per capita GDP, then levels off.

from no health care spending to $300 per person per year—less than a dollar a day—makes a substantial difference in life expectancy. Going from $300 to $2,000 makes almost no difference (see figure 3, page 69).

In fact, the largest increase in life expectancy occurs between $0 and about $100 per person per year in health spending. Additional spending increases life expectancy, but not as much as that first hundred dollars. Basically, the poor are getting more improvement in their health per dollar than the rich. Alternately, you can think of it in terms of the price of years of life expectancy. The first few years come cheap. The later years get progressively more and more expensive.

If enhancement technologies follow the same pattern, then what we'll see is not so much a gap between rich and poor as a time lag. Every new enhancement technology will initially be fairly expensive, and thus available only to the rich. In time, the technology will have competitors, or will go off patent, and its

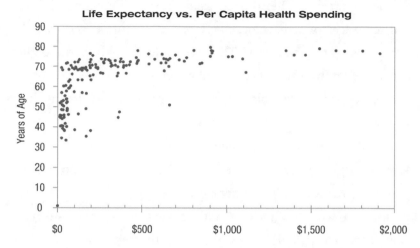

FIGURE 3 Life expectancy rises quickly with increases in per capita health spending, then levels off.

price will drop markedly. For any given type of enhancement, there will always be newer, more expensive techniques that work slightly better than the older, cheaper techniques. At any given time, the rich will be able to afford the latest techniques while the poor will not. But if the relationship between life expectancy and spending is any guide, the latest and greatest enhancements may cost anywhere from ten to one hundred times as much as older techniques while offering only a slight improvement. Both rich and poor will benefit, with the rich paying dearly to benefit first.

If, on the other hand, governments restrict access to enhancements—for example, by attempting to ban them—prices rise. Consider the cost of tobacco and marijuana. Both are easy to grow and widely used in the United States. There is nothing about the plants that should suggest a radical difference in price. Yet while tobacco sells for a few dollars an ounce, marijuana costs around $300 an ounce, or about *one hundred times* more.

Higher prices of illegal drugs may or may not be a bad thing. Higher prices of enhancements, however, would exacerbate the gap between rich and poor. The more effective a ban, the higher the price of a black market enhancement will be. As prices climb, the population able to benefit from an enhancement shifts toward the rich.

In the extreme case, in which a country is able to perfectly enforce a ban on an enhancement, the rich will simply go abroad. Attitudes toward enhancement vary considerably in different parts of the world. While the United States and other Western nations may ban genetic engineering or other enhancement techniques in the coming years, those techniques are likely to be legal in Asia and other parts of the world.

The richest Westerners will always have the ability to go abroad for their enhancements. A wealthy couple can choose to fly to China or some other location where an enhancement procedure is legal, stay for the necessary days, weeks, or months, and then return. If the procedure can be imported in the form of a drug or gene therapy, they're more likely to be able to pay the high cost of smuggling. A poor or middle-class couple doesn't have the same options. Prohibition would effectively put enhancements farther out of the reach of the poor. If equality of access is a goal, it makes more sense to encourage the development and use of enhancement techniques, to make them as widely accessible as possible, rather than restrict them.

Regulating Enhancement

An important step in spreading the power to alter our minds and bodies to more individuals is to reform the way enhancement technologies are regulated. A drug or gene therapy that enhanced memory or learning or slowed the aging process wouldn't necessarily be illegal today, but it would need to go through a regulatory process that would make it harder for doctors and consumers to get high-quality information about the safety and effectiveness of the technique.

In the United States, drugs, gene therapies, and medical devices are regulated by the FDA. Gene therapies are regulated essentially as drugs, and medical devices are regulated fairly similarly. To be approved for human use in the United States, any drug, gene therapy, or medical device must go through three stages of development.

First, preclinical research using chemical analysis, cell cultures, and animals must indicate that the drug is effective in treating a disease and has a high likelihood of being safe for humans. If this preclinical stage is successful, a researcher or pharmaceutical company can submit an "investigational new drug" application, or IND. If the FDA determines that animal and lab experiments have provided enough evidence of safety and effectiveness, the agency will approve the IND, and the drug moves on to human trials.

Human trials occur in three phases. Phase I studies are about safety. They use low doses in a small number of patients to look for any potential side effects. Phase II studies focus on effectiveness. They look at the impact of the drug on a larger population, usually a few hundred people, to determine if the drug is actually effective at treating the targeted disease or condition, and continuing to monitor for safety problems. Phase III studies look at effectiveness and safety in a yet larger population—several thousand people—over a period of many months to a few years.

Since the clinical phase involves at least three separate sets of human experiments, with time for analysis of the data between each one, the whole process takes a minimum of several years and at times as long as a decade. At the end of this period, if the drug is effective in treating the disease and safe in large populations for extended periods of time, the sponsor can submit a "new drug application," or NDA. If the FDA approves the NDA, the drug can be marketed and sold for the specific condition it was approved for.

Once a drug is on the market, physicians are allowed to prescribe it for "off-label" use—treatment of conditions other than the specific condition it was approved for—so long as that use falls within a nebulous definition of "the practice of medicine." As a practical matter, the FDA scarcely regulates physicians, so for most drugs, doctors are free to use their best judgment in prescribing it for related conditions. For example, a doctor would be effectively free to prescribe a drug approved for the treatment of Alzheimer's that also improved the memory of healthy young people to a college student cramming for an exam. The decision to do this would be up to the physician. The major exception to this general rule is any drug (typically those with recreational uses) defined as a controlled substance. Prescriptions of those drugs are monitored by the Drug Enforcement Administration.

While physicians are essentially free to prescribe drugs for off-label uses, pharmaceutical companies are much more restricted in what they can advertise. The FDA prohibits drug manufacturers from advertising any off-label use to either physicians or patients without the explicit permission of the agency, except in special cases. Since drugs can only be approved for treatment of disease, the "on-label" use will never mention any enhancement purpose, even if sound scientific research backs up the effectiveness of the drug as an enhancement.

It is interesting that the FDA's rules are reversed when it comes to dietary supplements. Labels and advertisements for dietary supplements may not make any medical claims. A vitamin bottle, for instance, can't carry the claim that it lowers the risk of cancer. However, dietary supplement packaging can carry virtually any other kind of claim, without any scientific basis. So an essentially untested herbal product can bear a label that claims "Improves Your Memory!" while a well-tested drug that has been scientifically shown to improve your memory cannot.

The FDA's lack of recognition of enhancement uses of drugs has two other effects. First, it reduces the incentive of a manufacturer to run the experiments that would determine whether a drug or gene therapy actually enhances those who use it. Since a pharmaceutical company can't market a product on this basis, there's a lower return on investment from any such studies.

Second, the safety trials required before approval of a drug or gene therapy typically focus on people who have the disease or condition being targeted. As a result, safety in people who *don't* have the disease isn't tested. A gene-therapy technique for muscular dystrophy won't be tested on athletes seeking to increase strength, and thus the safety of the product for use by athletes won't be known.

The net result of the current regulatory scheme is that consumers and doctors are denied useful information about the safety and effectiveness of potential enhancements. The regulatory model could be improved by doing two things, both of them quite simple.

First, the FDA and its analogous bodies in other countries should evaluate the effectiveness of drugs for enhancement claims as well as disease-treating claims. In most cases, the drugs that manufacturers wish to make enhancement claims for are also under review for medical uses. Testing and approval of a drug for an enhancement claim would be virtually identical to the current process, except that it would involve safety testing on a wider population and would also require verification that the drug actually enhances mental or physical function in the way its sponsors claim.

This small change to the FDA process would bring many beneficial effects. Doctors and physicians would now have reliable information about which drugs actually enhance human function. By default, they would also gain information regarding the effectiveness—or lack of it—of any drug that had *not* been approved by the FDA for an enhancement use. And we'd know more about whether drugs already being used as enhancements are in fact safe when used in that way. What's more, opening up the enhancement uses of a drug would increase total demand, driving down prices for both enhancement and therapeutic uses.

A second, slightly less important change we could make to the current regulatory scheme would be to restrict the enhancement claims that can be made for dietary supplements without research to back them up. This would help consumers distinguish between products for which there's scientific evidence of enhancement effects and those for which there's not. Practically speaking, this change is unlikely to happen in the near future. The FDA proposed a similar policy after the passage of the Dietary Supplement Health and Education Act in 1994, but lobbying pressure from the dietary supplement industry forced the agency to weaken the rules so that only "medical" claims are restricted.

Scientific testing of the safety and effectiveness of enhancement drugs and gene therapies would educate and protect consumers, and offer them a scientifically validated alternative to less-regulated products whose claims are rarely backed up by evidence.

If governments reform their regulatory processes to give doctors and consumers access to the best possible information about enhancement techniques, then market forces will probably suffice to bring the cost of enhancement technologies down to the level where most people can afford them. In general, governments are well served by allowing the market to do its job, thus leveraging the intelligence of millions of individuals who can choose the best product for the price, and reward the companies that bring them to market.

At times, though, the market does fail to bring prices down quickly enough to meet a pressing need. If this occurs with enhancement technologies, then governments should step in to reduce the cost.

Inequality in access to enhancement technologies brings the risk of stratification of the rich from the rest of the population. Some enhancements, like learning ability or memory, will increase earning ability. If the rich are able to buy these enhancements and the poor cannot, then the rich will be increasingly advantaged and the poor will fall even farther behind. For the rich, this would be a virtuous cycle of gains begetting more gains. For the poor, it would be a vicious cycle, as lack of access to enhancements prevented access to the best jobs, thus robbing them of the money they need to buy enhancements.

The recent history of technology has *not* followed this pattern. In real terms, the poor of the world are closing the gap with the rich—in life expectancy, in education, in access to technology. Like other technologies, biotech enhancements are likely to rapidly drop in price. The only reason to fear such a stratifying effect of biotech enhancements is their potential for rapid and dramatic effects. If biotech and neurotech enhancements increase earning power *faster* than the technologies drop in price, they could lead to runaway stratification. In this case, governments should act.

Governments should also be prepared to step in as a matter of opportunity rather than risk. As previous chapters have shown, enhancements that boost mental function can contribute to economic growth, and enhancements that slow aging or reduce disease can save society large sums of money that would otherwise be spent on health care. In the United States alone, every 1 percent increase in productivity adds $100 billion to the economy, and every 1 percent reduction in health-care costs saves $28 billion.

Given these impacts, biotech and neurotech enhancements can be seen as investments in valuable human capital. If enhancements are cheaply and widely available, then individuals and families will decide where and how much to invest in them. If for whatever reason the market fails to reduce costs to the point where most people can afford to purchase biotech augmentations of their minds or bodies, then governments have a tremendous amount to gain by investing their own funds in spreading access to mental and health enhancements to the public.

Government spending in these areas has precedents. England, Australia, Japan, and other countries already subsidize the cost of in vitro fertilization, with the effect that in Australia the procedure is four times more common than in the United States.

Governments also invest in enhancement in other ways. In the United States, federal, state, and local governments spent a combined $6,911 per student for elementary schooling in 2000, totaling nearly $382 billion per year. Over the lifetime of a student, that comes out to nearly $83,000 spent on public K–12 education, not considering the cost of college.

By contrast, most enhancement techniques would be considerably cheaper. The best comparison for gene-therapy costs, for instance, might be vaccinations. The per capita costs of influenza vaccines is around $15 in the United States today. In comparison to the money society already spends on education, biotech enhancements could provide a dramatic return on investment.

In the West, we view societies as being formed to secure the rights of individuals, and we embrace the freedom to make ourselves who we want to be. We work to increase equality of *opportunity*, so that as many individuals as possible might possess the power to steer and improve their own lives. History shows us that when individuals are given the power to make such choices, it serves to benefit society as a whole, resulting not in a stifling sameness but a rich diversity. It has also shown us that human beings are the ultimate resource, the most precious capital a society has. It makes good sense, on grounds of both principle and pragmatics, to invest in this resource. Rather than trying to restrict freedoms, governments ought to focus on empowering individuals and families, and reaping the benefits for all humanity.

CHAPTER 4

Methuselah's Genes

BENJAMIN FRANKLIN FAMOUSLY observed that nothing is certain in life except death and taxes. That hasn't stopped people from trying to avoid both, and while taxes have remained with us, we've made progress of a sort against death. Around the world, a child born in 2000 will live, on average, to the age of sixty-six, roughly twice as long as a child born in 1900 and three times as long as someone born a millennium ago (see figure 4, page 79).

Humans live longer, on average, than ever before. Yet until recently this progress has been made not by combating aging, but by protecting us against the things that kill us while we're young. Medicine has done very little to lengthen the lives of those who make it to old age. In 1900, an American male who had managed to survive to age seventy could expect to live to age seventy-nine. Today, a seventy-year-old American man will live on average to age eighty-two, a gain of just three years of life expectancy at that age over the last century—a trifle compared to the gains for newborns.

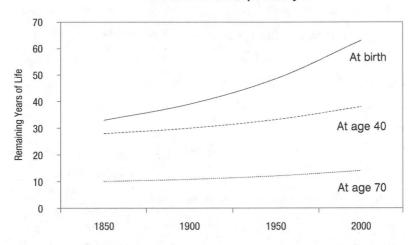

Worldwide Life Expectancy

FIGURE 4 Life expectancy at birth has gone up sharply over the last 150 years. Life expectancy at later ages has risen more slowly.

We've made such small progress in increasing life expectancy at more advanced ages because medicine has focused primarily on disease rather than on the aging process itself. Antibiotics have lengthened the average human life, but haven't changed the fact that our immune systems get weaker as we age. Heart surgery can repair acute damage but can't restore the heart to youthful health. Nothing we've done so far has addressed the fundamental issue: the older we get, the less capable and more vulnerable to health problems we become. If we are living longer, it's because more of us are living to be *old,* when what we really want is to *stay young.*

The vulnerability that comes with advanced age is captured in something called the Gompertz Law. Benjamin Gompertz was a self-educated Londoner living around the turn of the nineteenth century. His law states that in any species, the risk of death rises exponentially with age. Applied to humans, this means that our risk of dying from any cause roughly doubles every eight years. In essence, time wears down our minds and bodies. The diseases that kill us in old age, such as heart attacks and cancer, are largely symptoms of the aging of our bodies. Every time medicine cures a disease and lengthens life, humans run into the wall of another disease. Alzheimer's disease is an example—as the life spans of people in developed countries increase, more and more suffer from this disease.

This pattern is clear to demographer Jay Olshansky of the University of Chicago, who says, "If, indeed, we were to find a cure for cancer, life expectancy at birth would rise by about three and a half years. . . . And if we eliminated all cardiovascular diseases, diabetes and all forms of cancer combined, life expectancy at birth in humans would rise up to about 90." That's a roughly ten-year increase in life expectancy—a great advance, but still less than a third of the gain humanity made in the twentieth century.

Jonathan Swift's *Gulliver's Travels* contains a scene that puts our progress in lengthening human life in stark relief. It's the story of a race called the Luggnaggians and the immortals among them, who are called Struldbruggs. Most Luggnaggians are ordinary mortal creatures who die when their time comes. Struldbruggs are rare Luggnaggians whom nature has gifted (or cursed) with immortality. The Struldbruggs age but do not die. As time passes they grow ever weaker, frailer, and more miserable, but are never released from their lives.

This notion of a frail, ailment-ridden old age confirms our fears for our own last years—that we'll spend them hooked up to life-support machines, suffering through dozens of bodily ills, and mentally addled enough to require constant supervision. We all want to live, but not this way.

Critics of techniques to lengthen the human life span often

paint a Struldbruggian picture of our future. In these scenarios, a large elderly population is kept alive only through incredibly expensive medical techniques paid for by the young. And, in fact, current trends in health care and life expectancy provide some justification for such scenarios. As the population of the developed world ages, pension programs will come under pressure, medical costs will soar, and the number of young able-bodied and able-minded people available to pay for it all will shrink as a proportion of the population. This underscores the core point, that if we want to live longer and happier, we must begin addressing not only the symptoms of old age, but the phenomenon of aging itself.

Actually slowing aging—extending youth—would pay society substantial dividends. It would reduce the incidence of age-related diseases that account for roughly half of all deaths today. It would reduce the cost of health care across the world by hundreds of billions of dollars, as the young generally consume less medical care than the old. It would keep people mentally and physically fit longer, enabling them to pay their own way through society, rather than being a drain on the income of younger generations. In short, extending youth would address the very problems that longer lives are causing.

A pill or injection that prolonged youth would also be immensely popular. In 2002, American consumers alone spent an estimated $3.6 billion on putative "anti-aging" products ranging from antioxidants to growth hormone injections, despite the fact that not one of them has been proven to slow or reverse aging. Some of these products are sold by true believers. Just as often, they come from unscrupulous vendors who believe only in the money they can make selling questionable cures for aging to scientifically naïve customers.

The market for scientifically unproven anti-aging products has so disturbed real scientists in the field of gerontology that in 2002, fifty-one of them signed a joint communiqué titled "Position Statement on Human Aging." Their paper stated unambiguously that today, for human beings "there is no such thing as an anti-aging intervention."

The authors of the paper harshly critiqued existing "anti-aging" products on the market. Their goal was to educate a public that, in their eyes, is being duped by blatantly false advertising. Yet after debunking today's products, the authors ended on a more optimistic note, stating, "Our rapidly expanding scientific knowledge holds the promise that means may eventually be discovered to slow the rate of aging." In the pages that follow we'll talk about some of those promising discoveries.

The Future of Mortality

"In the past few years a number of experiments have completely blown away the theory that aging could not be slowed," says biologist Gordon Lithgow of the Buck Institute for Age Research in Novato, California. Lithgow should know. He's one of a vanguard of scientists who've discovered techniques that can actually slow aging in animals and that one day may do the same in humans.

Andrzej Bartke, one of the signatories of the 2002 position paper, also knows firsthand the potential of science to slow aging and extend life. He's seen it in the mice he works with in his lab at Southern Illinois University. On January 8, 2003, one of his genetically modified mice died just a week short of its fifth birthday. Unenhanced lab mice die at about half that age; the very longest lived lab mice live for three years. Bartke's genetically enhanced mouse had a life span 66 percent longer than the maximum life span of unenhanced individuals of the species—as if a human had lived to be two hundred. Furthermore, the mouse continued in good health right up to the end, free of the diseases and fragility that plague most old mice.

In the late 1980s, the work of geneticist Tom Johnson at the University of Colorado broke open the field of genetic age interventions. Johnson started working on the genetics of aging in 1979. At the time he, like most other researchers, believed that aging was such a complex process that extending the life span of an organism would require changes to hundreds or even thousands of separate genes. Then, in 1983, his colleague Mike Klass found several mutant strains of a tiny nematode (a phylum of parasitic worm) that had exceptionally long life spans—up to twice as long as the norm for their species (*C. elegans*).

After Klass left academia for industry, Johnson took over his work. Initially he assumed that to have a life span twice as long as normal, the worms must have had mutations to a large number of genes. He gave the project to David Friedman, an undergraduate then working under him. What Friedman found was that, contrary to what anyone suspected, out of the nineteen thousand genes in the creature's genome, the long-lived nematodes had a mutation in just a single gene (later named age-1). In papers in 1988 and 1990, published in *Genetics* and the prestigious journal *Science,* Johnson and Friedman laid out their amazing findings: a change to a single gene could double the life span of the nematode. Later research showed that mutations of two genes combined could even triple the worm's life span.

Another leading researcher in the field of aging, University of Idaho evolutionary biologist Steven Austad, underscores just how tiny a genetic change this is: "Worms have 100 million nucleotides in their genome approximately. If you change one of those, one of the letters, one of the 100 million letters of the genetic alphabet, you get a doubling of life span. If you change two of the letters, if you change one letter in one gene, one letter in another gene, you'll get a tripling of life span."

While Johnson and Friedman's research opened the way for much of the field of life extension, it wasn't well received at first. "My initial findings," Johnson says, "were pretty much ignored by most people. . . . I really felt ostracized. I remember very much a comment by a colleague once to one of my students to the effect that 'You are so smart, what are you doing working with that charlatan? Someday the world is going to find out that he has been faking data.' "

In the end, the scientific method won out over dogma. Five years after Johnson's publication of his results, an established researcher, Cynthia Kenyon at the University of California at San Francisco, discovered that another gene, called daf-2, could also lengthen the life of these worms. With two different labs showing similar results with different genes, it was harder to claim that the life extension was an error or a hoax. Prominent scientists started to take notice of this idea that small genetic changes could extend life, and set about searching for such genes themselves. The field of genetically extending life was born.

Since then, scientists have discovered dozens of genes that can increase life span in a wide variety of creatures. Some of them extend life by just a small increment, while others can more than double life span. Scientists now know of more than a hundred genes that can slow the aging process in nematodes, fruit flies, worms, and mice, and more are being discovered every day. These genes that researchers are altering aren't just present in animals—they're also present in us humans. This suggests that making the same genetic changes in men and women might slow human aging. A host of researchers in both academia and private industry are working on ways to use this genetic knowledge to design drugs that might extend our lives by decades or more.

Life extension genes fall into a handful of different families. Genes of the earliest discovered family have variants that reduce the body's sensitivity to insulin or a related chemical called insulin-like growth factor. A second and more recently discovered family of genes works by protecting cells against the ravages of free radicals, highly reactive chemicals that bounce around causing damage to whatever they touch.

Age-1, the gene that Johnson discovered, is in the insulin-regulating group, which is today the best-studied family of life extension genes. Genes of this sort show up in almost every creature alive. Scientists have lengthened the lives of nematodes, fruit flies, and mice by making changes to these genes.

Insulin and its cousin, IGF-1, are messengers the body uses to send signals from one part to another. For example, when a muscle cell is contacted by a molecule of insulin, the muscle cell responds by sucking up more sugar from the bloodstream and storing it away as a chemical called glycogen for future use as fuel. Age-1 and its related genes are involved in making the receptors on each cell that pick up these chemical messages. Johnson, Kenyon, and other researchers have extended the lives of nematodes by reducing the function of age-1 and related genes, making the animal's cells less sensitive to insulin.

These genetic changes not only lengthened the worms' life span but also made the worms more resistant to stresses of all sorts. Between 1993 and 1995, the biologists Simson Melov and Gordon Lithgow at the Buck Institute for Age Research ran a series of stress experiments on nematodes with mutated age-1 and daf-2 genes. They put the worms in high temperatures that would kill normal creatures and exposed them to toxins that they shouldn't be able to survive. In both cases, they found that the worms that had been genetically modified for longer life were also more resistant to the elevated external stresses—making the genes even more attractive as targets for life-extending drugs in humans. After all, no one wants to be old and frail, but a single treatment that could keep you young and also protect you against toxins and other stresses sounds pretty appealing.

Exactly how and why mutations of the insulin pathway slow aging remains something of a mystery, but it may be related to the protection against stress that Melov and Lithgow discovered. In a review published in *Nature,* researchers Toren Finkel and Nikki Holbrook of the National Institutes of Health pointed out that virtually every age-slowing genetic change discovered to date confers added resistance to stresses like heat and toxins that could normally damage DNA. By contrast, virtually every genetic mutation that accelerates aging also makes creatures more vulnerable to heat, ultraviolet light, toxins, or other environmental stresses that can damage their genes. If Finkel and Holbrook's hypothesis is correct, the insulin-like pathway in the body plays a role in protecting our cells against genetic damage, and the changes to this pathway that slow aging increase its protective function.

Some life-extended organisms do suffer from other side effects, though. Kenyon's daf-2 mutants experienced reduced fertility and other problems that seemed tied to their early development. By taking genetically normal nematodes and reducing the function of daf-2 only in adulthood, Kenyon was able to bring them the benefits of extending their lives while avoiding the unpleasant side effects.

It's fine to extend a nematode's life—but is this relevant to humans? More important than how these genes work in nematodes is how they work in mice. Mice are mammals, more similar in their biology to humans than insects or worms, so we'll learn far more about the prospects of extending human life by studying them than nematodes or fruit flies. The quest to lengthen the lives of mice by reducing their sensitivity to insulin or IGF-1 has been tricky. In 2000 and 2001, a number of studies showed that it could be done, but at a price. Four different labs produced "dwarf" strains of mice that either lacked or could not respond to growth hormone, which affects the body partly by triggering the release of insulin-like growth factor 1. By blocking the effects of growth hormone, the researchers indirectly blocked the effects of IGF-1. The modified mouse strains lived 30 to 60 percent longer than normal mice, but unfortunately, the mice all experienced serious side effects. Growth hormone is essential to development; lacking it, the mice ended up physically stunted—around half the size of other mice—and two of the strains are almost completely infertile.

While the side effects of reducing growth hormone function were a blow to the study of the insulin and IGF-1 anti-aging techniques in mice, they didn't kill the idea of modifying IGF-1—related genes to extend life. The first four studies all made larger changes to the mouse's genes than anyone wanted. The aim was not to disable growth hormone—reducing growth hormone was just a handy way to reduce the function of IGF-1. So, was it possible to make a more precise change to the mice? Could researchers reduce the function of the IGF-1 receptor without knocking out some other important part of the mouse's biology? If so, would that very targeted change extend life? And would it have major side effects?

Two partial answers came near the end of 2002 and the beginning of 2003. In late 2002, a French geneticist, Martin Holzenberger, reported that his team at the National Institute of Health and Medicine in Paris had created mice that had only one copy of the gene for the IGF-1 receptor, instead of the normal two copies. Up until then no one had succeeded in doing anything so specific. Holzenberger's mice live longer than his controls—16 percent longer in the case of the males and 33 percent longer in the case of the females. They're also full size, eat as much as normal mice, and seem just as active. And they're more resistant to stress—they survive exposure to a toxin called paraquat hours longer than control mice.

Holzenberger's study had a few problems, though. The strain of mice he started with, before his genetic modifications, wasn't very long-lived, especially in the rather crowded conditions in which the mice were housed in this experiment. As a result, his life-extended mutant mice, which lived longer than the strain he started with, still died earlier on average than other strains of mice kept in better environments. Still, Holzenberger clearly showed that by modifying the IGF-1 gene he could lengthen the life of one strain under one set of conditions. Combined with the evidence in nematodes and other strains of mice, this suggests that IGF-1 will lengthen the lives of mice living in better conditions as well.

A related study showed that mice that lacked an insulin receptor in their fat cells lived longer. Researchers at Harvard increased the average life span of mice by about 18 percent. The mice looked and acted normal. They had less body fat but were otherwise normal size. And they ate about one and a half times as much as other mice of the same size but remained thin and healthy, and still outlived the genetically unmodified mice.

All told, changes to the family of genes that controls the insulin and insulin-like receptors have extended life in every species they've been tried in. Researchers have even found that they lengthen the life of yeast. Now what everyone wants to know is, will such changes work in humans?

There are two reasons to believe that these genes could be used to increase human life span as well. First, age-1, daf-2, and other life-extending genes that we haven't talked about are very similar to genes that humans carry for the insulin and IGF-1 receptors. So the molecular switches we've flipped in nematodes, flies, and mice exist in our genetic code as well.

Second, it's telling that IGF-1–related changes have increased the life span of mice, fruit flies, nematodes, and yeast—incredibly different species. Mice are relatively close relatives of ours, having diverged from us on the evolutionary tree around 75 million years ago. Nematodes, fruit flies, and yeast, on the other hand, diverged from us hundreds of millions of years ago—more than a billion in the case of yeast. Yet the same life-extension techniques seem to work on all of them.

This is a profound discovery. The same techniques extend the lives of warm-blooded creatures, cold-blooded creatures, and creatures without any blood at all. They work on species that reproduce by budding, species like nematodes where every member is a hermaphrodite, species that reproduce by laying eggs, and species that have sex and nurse their embryos in the womb, like us. The same techniques lengthen the lives of nematodes, which possess less than a thousand cells, and mice, made of more than a trillion cells. They slow the aging of creatures we diverged from as long ago as a billion years and as recently as 75 million years ago. They prolong the youth of animals with brains with billions of neurons and creatures with no brains at all. Given how different these species are, the IGF-1 gene changes must be affecting aging at a very fundamental level.

Cynthia Kenyon, for one, believes that this broad spectrum of species provides the crucial evidence that the IGF-1 gene can be used to extend human life as well. Holzenberger's work in mice is "wonderful," she says. "It shows quite clearly that [the insulin-related longevity pathway] is conserved in mammals, and that's really powerful. If it's true in worms and flies and mice, the chance that it's not true in humans is small."

Kenyon is confident enough to bet her time and her career on it. In 1999, she and prominent researcher of aging Lenny Guarente cofounded Elixir Pharmaceuticals. Years earlier, Guarente had discovered that a gene called SIR2 could lengthen life in yeasts and nematodes, apparently by blocking the activity of harmful genes inside the body. Humans and other mammals carry descendants of these genes called sirtuins, and researchers are currently looking at lengthening the lives of mice by modifying these genes.

Between them Guarente and Kenyon have more than 250 scientific publications to their names. Along with Tom Johnson, they're among the most respected researchers in the field of genetically modifying animals to increase their life spans.

In late 2002, Elixir merged with Centagenetix, a company looking at the genes of exceptionally long-lived humans to find drug targets to cure age-related diseases. They've raised $20 million to

$30 million to fund their research. Elixir's strategy for finding drugs that lengthen life is simple. Since making cells less sensitive to IGF-1 extends life in so many animals, they'll look for drugs that can do the same thing artificially.

The IGF-1 receptor, like all other receptors, is a bit like a three-dimensional puzzle piece. A molecule of IGF-1 fits snugly into the pocket of the IGF-1 receptor. This tells the receptor that a signal has been received, and the receptor in turn notifies the rest of the cell. One way to reduce the sensitivity of the receptor is to find a molecule that makes it harder for an IGF-1 molecule to fit snugly into the right place. Imagine inserting an odd-shaped piece of cardboard inside the pocket that is the IGF-1 receptor. The foreign piece of cardboard would make it harder for the other puzzle piece (the molecule of IGF-1) to "dock" in the right place. A molecule that interferes like this is called an antagonist because it works against the function of the receptor.

Elixir Pharmaceuticals is in the process of searching for compounds that might work this way. After they find a good target, they'll test it in lab animals. If the drug extends life, and if it's safe, then they'll slowly and carefully move to tests on humans. The whole process, if it succeeds, will take ten years or more. Of course, the Food and Drug Administration and other regulatory agencies only approve drugs to cure diseases, and they don't (yet) consider aging a disease. So the most likely strategy for Elixir is to show that their drug helps stave off some symptom of aging, like cancer or diabetes or heart disease. Although such a drug wouldn't be advertised as slowing aging, some people would probably seek it out as preventive medicine. If history is a guide, they'd find sympathetic doctors to prescribe it—just as motivated patients have a fairly easy time acquiring Viagra, antidepressants, or other prescription drugs from their physicians for off-label use.

To prevent patient harm, regulatory bodies will need to insist on tests of these drugs designed to show how they affect perfectly healthy people not yet suffering from cancer, diabetes, or heart disease. For now, the scientific and business communities aren't worrying about the regulatory issues. If drugs can be found to slow human aging, there will undoubtedly be a huge market for them. Even if Elixir isn't the company to find such a drug, the search is on. Around the world, researchers of aging now know that variants of genes related to IGF-1 extend life in a variety of creatures. Other labs have identified additional genes in this family, and pharmaceutical companies have their eyes on the field.

Free Radicals

The other large and growing family of genes that extend life are those that protect cells against free radicals. Free radicals are molecules inside every cell of your body that are electrically charged in such a way that they can easily chemically react with other parts of the cell. In particular, they have a high affinity for electrons—they will frequently bump into another molecule and steal an electron from it. When a free radical bumps against a strand of DNA and strips off an electron, it can cause a mutation. Over time, these mutations build up, causing your cells to malfunction in subtle and then progressively less-subtle ways.

Your cells do an amazing job of repairing the damage that free radicals cause, but occasionally something gets past their self-repair machinery. Much of the damage from free radicals accumulates in the power plants inside your cells called mitochondria. As mitochondria produce the energy for the rest of your cell, they create a large number of free radicals as by-products, which often damage the small amount of DNA in the mitochondria themselves. As damage accumulates in these cellular powerhouses, they're less able to produce energy for the rest of the cell; and they—and the cell—just can't work as hard. Muscles become weaker. Brain cells become slower. Some cells just stop working completely.

Researchers have uncovered the effects of free radicals and mitochondrial DNA damage in experiment after experiment. Different species of animals produce different amounts of free radicals. Longer-lived species produce fewer free radicals and shorter-lived species produce more free radicals. Researchers have also created tests to measure the amount of damage that free radicals do. In creatures with short life spans, they have found lots of free-floating proteins and other molecules that have been damaged by free radicals. In longer-lived species, the concentration of free radical–damaged molecules is much lower.

Scientists studying nematodes have also found that older nematodes have more damaged mitochondrial DNA than younger nematodes. Life extension genes like age-1 slow the rate of this damage—in fact, most of the life-extension genes we've talked about seem to slow the rate of free radical damage and make creatures more resistant to toxins that generate free radicals.

Most important, genes that fight the effects of free radicals can also slow aging. Antioxidants are chemicals that stabilize free radicals and make them harmless. Vitamin C is a popular antioxidant. Unfortunately, experiments have not shown that feeding animals vitamins can lengthen their lives, but experiments that have increased the body's supply of its own antioxidants have done wonders.

Inside the body, two of the most important antioxidants are chemicals called superoxide dismutase (SOD) and catalase. In 2002, scientists created two strains of fruit flies that produce more SOD. These strains live about 40 percent longer than normal fruit flies. In addition, a genetically modified strain of fruit fly, which creates extra methionine sulfoxide reductase A (another important antioxidant in the body), also lives substantially longer than normal fruit flies. A fourth strain of fruit flies carries a gene, whimsically named methuselah (for the long-lived Old Testament patriarch), that gives it extra protection against free radicals. These flies live around 33 percent longer than normal and are especially resistant to heat, stress, and starvation.

Removing a gene called p66shc seems to make the cells of mice more resistant to free radicals, and also lengthens life span by about 30 percent with no obvious side effects. The University of Idaho's Steve Austad is collaborating with a lab that has lengthened the lives of another strain of mice by genetically engineering them to produce more of the antioxidant catalase, specifically inside their mitochondria.

Most important, there are at least two drugs that have now been found to lengthen the lives of animals, and which seem to have antioxidant properties. Simon Melov and Gordon Lithgow found that two drugs that mimic the effects of antioxidants inside the body could lengthen the lives of nematodes. Manufactured by a private company called Eukaryon, EUK-8 and EUK-134 act inside cells like SOD and catalase. When Lithgow and Melov fed EUK-8 and EUK-134 to nematodes, they lived on average 44 percent longer than controls. Currently Lithgow and Melov are testing these drugs on mice as well. Another drug, 4-phenyl butyrate, protects fruit flies against the effects of free radicals and also lengthens their lives.

In summary, a large and growing body of evidence suggests that increasing the body's defenses against free radicals can extend life. In the few years since the first evidence that life span could be increased this way appeared, in 1999, researchers have made impressive progress. Their successes are sure to attract more researchers to the field. If the pace of discovery continues, we may see therapies to increase the human life span enter human trials within the next decade.

CHAPTER 5

Choosing Our Life Spans

WHILE GENETICISTS HAVE been looking at the use of genetic engineering to prolong youth, another group of researchers have found a simple and reliable way to lengthen life: feed animals less food while maintaining their full intake of vitamins and other essential nutrients. This is what scientists call caloric restriction, or CR. For reasons we've just recently begun to understand, CR lengthens the lives of mice, rats, nematodes, fruit flies, spiders, guppies, dogs, chickens, and plenty of other species. There's preliminary evidence that it works in humans as well. When started at childhood in animals, it usually lengthens life by 30 to 40 percent. For a typical American, that would add up to thirty years to life and would mean an average life expectancy of 107 instead of 77 and a maximum life span of about 170 instead of 120.

Most of us already know that keeping caloric intake at reasonable levels is good for us. Nevertheless, obesity is on the rise in the developed world, especially the United States. Caloric restriction has far greater benefits than just eating a reasonable diet. No reasonable diet will extend your life 40 percent beyond the human norm. No reasonable diet will keep you healthy and fit decades longer than your free-feeding peers. But caloric restriction is also much more severe than a reasonable diet. Imagine *always* choosing the salad instead of the double bacon cheeseburger or never eating dessert for the rest of your life. It requires not just discipline, but intensive planning of your meals. It forces you to pay careful attention to your calorie and nutrient intake. Most of us just don't have the will or the time put ourselves on a CR diet.

Scientists researching CR know full well that most people would not submit to such a regime. Humans enjoy their food. Scientists have no illusions about human willpower. But they also know that humans will buy products that keep them young or that *keep them thin*. Inspired by this opportunity, some prominent and reputable caloric restriction researchers are now searching for caloric restriction *mimetics*—drugs that mimic the effects of CR on the body, while still letting you eat what you want.

Researchers have known for almost a century that reducing the caloric intake of lab animals can have health benefits. In 1914, Francis Rous (best known for winning the Nobel Prize in 1966 for his work on cancer) showed that reducing the food intake of rodents substantially reduced their incidence of tumors. Over the following decades, researchers slowly expanded this picture. In 1930, Clive McCay and his colleagues at Cornell University showed that rats that had their food restricted lived longer than rats that ate all they wanted. By the 1960s, the case for caloric restriction in animals was solid. Scientists all over the world had reproduced CR's life-extending effects on several different animals.

Scientists have now studied CR on a wide variety of other

creatures, including rats, fish, cows, and dogs. In almost all of these species, caloric restriction started in childhood extends life by 30 to 40 percent. Starting it by early middle age extends life by about 20 percent. The numbers vary from species to species, but most fall inside this range.

Exactly how and why caloric restriction works is not yet clear. Increasingly, however, researchers believe that it affects the same mechanisms as genetic changes that slow aging. In particular, organisms that consume less food seem to experience a slower rate of damage to the DNA inside their cells. This may be a result of producing fewer free radicals—a cell that is provided less energy will produce fewer of the waste products that damage it. Or it may be for other reasons not yet discovered. Either way, caloric restriction does more than make animals live longer—it seems to keep them young and healthy longer too.

Calorically restricted mice are more physically active than their free-feeding peers and maintain that activity later into life. Evolutionary biologist Steve Austad says, "One of the striking things that you don't often read about these calorically restricted animals is they have enormously increased spontaneous levels of activity; whereas a young mouse might run a kilometer in a night, a caloric restricted mouse might run six or seven kilometers in a night. Whereas, a normal mouse would stop running at all by the age of eight months, these animals are still running several kilometers a night at the age of two years."

Caloric restriction also protects the mind against the ravages of age. After adulthood, lab animals show a slow decline in memory and other mental functions, just like humans. Lab animals on CR diets still show this decline, but it's remarkably slowed. Studies have shown that old mice on CR diets do better on memory and other mental tests than mice of the same age that have been allowed to eat whatever they want. In 2002, Christiaan Leeuwenburgh at the University of Florida showed that caloric restriction slows the rate of death of neurons in the brain, possibly explaining why CR animals stay mentally fit.

CR staves off disease as well. Calorically restricted animals have a lower risk of cancer, heart disease, diabetes, and just about every other chronic life-threatening disease. CR animals are also more resistant to stress. They can survive toxin levels, heat levels, and surgeries that other animals can't. Their immune systems decline more slowly than do the immune systems of free-feeding animals.

Calorically restricted animals *look* younger too, outside and in. They stay sleek and trim. Their hair stays where it's supposed to and retains the color it had in youth. Inside they have the biochemistry of younger animals. As animals age, the levels of glucose and insulin in their blood rise, and their levels of a hormone called DHEAS (dehydroepiandrosterone) decline. Old calorically restricted animals have lower blood glucose and insulin than normally fed animals, and higher DHEAS.

One of the most interesting effects of caloric restriction is what researchers call compressed morbidity—a shortening of the period of ill health at the end of life. CR animals often seem to live their lives in good health, free from major diseases, and then just keel over one day, long after their free-feeding peers have all died. If CR-based techniques work for humans, this phenomenon could have profound implications for human society. Rather than a long period of infirmity at the end of life, with its attendant medical costs and burdens on loved ones, life-extended humans might more frequently die in the prime of health, while still living as independent and active members of society.

These results have been shown in a wide variety of animals. Just recently, scientists have started to find similar effects in primates, our closest relatives. Since 1987, the National Institute of Aging (NIA) in Bethesda, Maryland, has been studying the effect of caloric restriction on a colony of rhesus monkeys under the supervision of researchers George Roth and Mark Lane. In 1989, the University of Wisconsin at Madison started a similar study. All told, there are about three hundred primates being studied today between the two institutions. In addition to rhesus monkeys, the NIA is also studying caloric restriction on shorter-lived squirrel monkeys.

Rhesus monkeys live on average twenty-four years in the lab, with a maximum life span of forty years—much shorter than a human life, but much longer than the three years that mice are given. Because rhesus monkeys live so long, though, it's going to take a long time to see the full effects that caloric restriction has on these animals.

In the less than two decades that have passed since the study began, though, the researchers *have* seen marked effects. In all of these cases, the CR monkeys are compared to a control group that can eat whatever they want. Roth and Lane have found that the CR monkeys have lower insulin and blood glucose, just as we'd expect. They have higher levels of DHEAS, a hormone whose level normally declines with age. They have lower blood pressure and lower cholesterol. They have more flexible arteries, like younger animals. About half as many as the free-feeding monkeys show any signs of heart disease, cancer, or diabetes. Most important, the death rate so far in the CR group has been just over half of the death rate in the free-feeding group. While the total number of deaths in the colony is too small to be statistically significant, the lower rate of death is also consistent with all the other findings. In short, caloric restriction seems to be keeping these primates young.

Roth and Lane have also shown that some of the biochemical changes that come with caloric restriction in animals are associated with longer life in humans. The Baltimore Longitudinal Study of Aging (BLSA), the most comprehensive study of aging ever conducted in humans, provides benchmarks for comparison. Calorie-restricted monkeys have lower body temperature, lower insulin levels, and higher levels of DHEAS for their age than the norm. Each of these traits is associated with longevity in humans in the BLSA. In addition, humans who practice caloric restriction show similar biochemical changes, suggesting that CR affects humans the same way it affects other animals. Put it all together and there's a good case to be made that caloric restriction is likely to work in humans.

There are drawbacks to CR as well. Calorically restricted animals have less fat to insulate them and have lower core body temperatures, making them more vulnerable to cold. CR animals are less fertile and less interested in mating than non-CR animals. They're smaller and less muscular than animals that eat all they want. And they may have weaker immune systems when they're young (though their immune systems stay healthy longer).

Humans who have voluntarily tried caloric restriction tell similar tales. Almost all of them report a loss of libido. Most are cold at temperatures that others find comfortable. Some report bouts of shivering. And many say that they fantasize about food constantly. Is a longer life worth it, if it comes with side effects like these?

Mimicking Caloric Restriction

If CR does work in humans and even if we can somehow elimi-nate the side effects, it may still remain unpopular. Most of us just aren't going to subject ourselves to the lifelong denial of the plen-tiful food that surrounds us. So to bring its benefits to humans, re-searchers will need to develop drugs that mimic the effects of CR while allowing us to eat all we want. These drugs will also need to avoid the side effects of CR. Achieving that combination in a sin-gle drug will be challenging, but there's reason to believe it's doable.

In the film *Monty Python and the Holy Grail,* there's a scene where an old man is thrown onto a cart of dead plague victims. "I'm not dead yet!" cries the still-living geezer, but his being alive is an inconvenience to those who just want him to hurry up and die. In 2000, Blanka Rogina and Stephen Helfand, biologists at the University of Connecticut, had the same problem with fruit flies. That year, Rogina and Helfand were running an experiment on genetically modified fruit flies. The experiment had nothing to do with aging or life extension, but as time passed, some of the flies just weren't dying when they were supposed to. The researchers were impatient—all the flies had to die before they could finish the experiment. Then the realization hit—they'd found a gene that was extending the life of these flies. Further research showed that to be the case.

Helfand and Rogina named the gene INDY, short for "I'm not dead yet," in honor of the old man in the Monty Python flick. Flies that carry a specific alteration of the INDY gene have a flaw in their metabolism. They can't quite get all the energy out of the food they consume, and this puts them in a state similar to caloric restriction. INDY flies live on average almost twice as long as other flies, and as far as the researchers can tell, the INDY flies are just as active, just as large, and just as healthy as normal flies.

So far the researchers have found only one disadvantage to the INDY flies. When they're fed all they want, the flies are more fertile than their unmodified cousins. But when food is scarce, they're noticeably *less* fertile. This makes sense—because their metabolism is inefficient, they need to eat more to stay alive and healthy. Conversely, they're more sensitive to starvation than flies that get more energy out of each morsel they eat.

The finding of the INDY fly points to a possible mechanism for a life-extension drug for humans. Humans carry a gene very similar to INDY that's involved in metabolism; a drug that targets this gene might cause the same effects as those seen in the INDY flies.

Rogina has also collaborated with geneticist Stewart Frankel at Yale. In a study related to the INDY gene, Frankel and Rogina found that flies with a genetic mutation that reduced the level of an enzyme called Rpd3 lived 33 to 50 percent longer than normal flies. Like INDY, Rpd3 is involved in metabolism. More important, a drug that researchers think affects Rpd3 has already been shown to lengthen the lives of fruit flies. In 2002, Hyung-Lyun Kang at the National Institutes of Health showed that feeding fruit flies a drug called 4-phenyl butyrate extended their lives by about 40 percent. These examples raise the possibility that in humans, a drug, not just gene therapy or caloric restriction, could be used to slow aging.

At least three other teams are currently working on drugs that mimic the effects of caloric restriction. Roth, Lane, and the neurobiologist Donald Ingram published an article in 2002 in *Scientific American* describing their search for a drug that extends life by partially blocking the use of sugar inside the cell. They're currently working with a drug called 2-deoxy-D-glucose (2DG). 2DG is too dangerous for human use—it becomes toxic if too much is taken—but at low, carefully measured doses it may have life-extending effects. The researchers are currently feeding 2DG to laboratory rats, and the rats are showing some of the signs of caloric restriction—lower blood glucose, lower insulin, lower body temperature, and slightly reduced weight—even though they eat all they want. In another year or so, we'll know whether these rats actually lived longer than the controls. If the rats *do* live longer, then we'll have evidence that a drug can mimic the effects of caloric restriction. Roth has formed a company called GeroTech for the purpose of bringing these products to the market.

And he's not the only one. "There are another dozen companies working on CR mimetics, most of them using different mechanisms," he says. One of those companies is Boston-based Biomarker Pharma, founded by biologist Stephen Spindler. Biomarker's approach is to find the genes that become more active in animals that are placed on a CR diet and use these as targets for drugs. Spindler and his colleagues at Biomarker are studying the genes that are expressed differently in animals on caloric restriction diets—the genes that become more or less active when an animal goes on CR. They've already found that most of these gene expression changes happen very quickly, within just a few weeks of starting a CR diet. Using this information, they're looking for drugs that mimic the effects of these genes on the body. Basically, they have a systematic way to find new drugs like 4-phenyl butyrate and 2DG.

So we've seen that caloric restriction works in almost every animal it's been tried in. It seems to be working in monkeys, our close relatives. And various studies in humans suggest that it will probably, though not certainly, extend human life as well. Animals with genetic mutations that reduce the amount of energy they can get out of food seem to live longer, but without obvious signs of the side effects that plague CR animals. Researchers have already found one drug that works on one of these genes and extends life. A second drug is producing physical changes similar to CR, and we'll soon know whether it extends the life of rats. Using gene-expression studies, scientists are likely to produce many more of these drugs in the coming years. We can't be sure we'll ever find a drug that mimics CR and is safe for humans to use, but today the evidence suggests that it's possible.

The life-extension techniques we've talked about all turn out to be related to each other in some way. Calorically restricted animals and animals with mutations to their IGF-1 receptors both seem to suffer less damage from free radicals. Caloric restriction, IGF-1–related genetic changes, and antioxidant-related genetic changes all render animals more resistant to toxins, ultraviolet light, heat, and other stresses that can increase the number of free radicals in the body. And animals on caloric restriction diets have less IGF-1 in their bloodstreams.

It's also important to reiterate that all three of these techniques seem to preserve healthy, youthful life. Animals that have their lives extended by any of these techniques stay physically active to greater age than their peers who go untreated. They retain their ability to learn and remember to older ages than normal animals. They develop fewer tumors and less heart disease. In old age they look and act like younger animals. When they finally die, it's often without any signs of the cancers or stiff arteries or atrophied muscles that plague other old animals. As far as we can tell, all three of these techniques actually stretch out the prime of these animals' lives.

Despite these common threads, the three techniques have important differences and can complement one another. In fact, studies that have combined caloric restriction with mutations in the IGF-1 receptor gene have already increased life span more than either method alone. Andrzej Bartke, at Southern Illinois University, showed that Ames dwarf mice, a breed of long-lived mice that lack growth hormone, live even longer when placed on caloric restriction. Ames dwarf mice live about 50 percent longer than normal when they can eat all they want. When placed on a caloric restriction diet, they live around 75 percent longer than normal. Researchers have found similar effects when they've put fruit flies with a mutation in their insulin-like receptor gene on caloric restriction (see figure 5). Seventy-five percent life extension in a mammal is fairly incredible. If we could increase human life expectancy by the same amount, the average human life span in the developed world would reach 135 years, and the maximum human life span would surpass 200 years.

Thus, the IGF-1 receptor techniques and caloric restriction

Percentage Increase in Mouse Life Span

FIGURE 5 Caloric restriction and Ames dwarf genetic mutations both extend mouse life span. When combined, they have an even greater effect.

work well together. What about the antioxidant techniques that guard against free radical damage? Increasing life span by boosting resistance to free radicals is a new field, and so far no one has published any results from experiments that combine free radical resistance genes with caloric restriction or insulin-like receptor mutations. At least one research group is planning such an experiment, though. In the coming years, we'll find out whether all of these techniques can be combined to slow aging even further and target even more striking results.

A Cure for Old Age?

The approaches to slowing the aging process we've talked about so far are based on experimental successes. They all take the basic approach of finding a technique that slows down aging in lab animals and then making drugs that mimic that mechanism. Gerontologist Aubrey de Grey, of Cambridge University, is one of the more controversial figures in the field of aging; he believes that there are more effective techniques that could dramatically lengthen human life and asserts, "Indefinite postponement of aging may be within sight."

This is an ambitious statement, yet de Grey is a respected thinker in the field. In 2003, he chaired the annual meeting of the Institute of Applied Biogerontology, one of the more prestigious of the international aging conferences. He's also one of the signatories of the position paper that dismisses all current "anti-aging" products on the market as scams. De Grey's optimism may stem from his unique viewpoint on the field. Originally trained as a computer scientist, he approaches the basic problems of aging more like an engineer than a biologist.

In a paper coauthored with respected aging researchers Bruce Ames, Andrzej Bartke, Judith Campisi, and others, de Grey laid out a multiprong strategy that he believed would all but eliminate aging in humans. What's more, the same techniques could restore us to youthful vigor.

The most widely acknowledged of the problems de Grey focuses on is one we've already discussed—the slow decay of the mitochondrial powerhouses in our cells, thanks to the constant barrage of free radicals. Mitochondria are particularly susceptible to damage because they carry their own DNA, but they lack the powerful defenses and repair abilities that they need to protect it.

99.95 percent of the DNA in each cell in your body is stored in the cell nucleus. The nucleus of a cell is a safe haven for DNA. Free radicals and other dangerous molecules are mostly kept out of it by a protective membrane. Inside the nucleus, repair enzymes are constantly checking your DNA for problems and correcting them. Your mitochondria, on the other hand, reside outside the nucleus along with the rest of the cell's machinery. Most of the proteins needed to build mitochondria are located in the cell nucleus. Evolution has moved them there over time. But the job isn't quite done. The last thirteen genes that mitochondria need (out of about one thousand) are in the mitochondria. Since the mitochondria lack the defenses and repair mechanisms of the nucleus, this DNA is quite vulnerable to damage from free radicals and other sources.

De Grey proposes that we finish the job that evolution has started: moving the mitochondrial DNA into the cell nucleus, where it will be better protected against aging. The way to do this is by inserting the last thirteen genes involved in creating mitochondria into our nuclear DNA. As de Grey points out, researchers have already moved one of those genes into the safe harbor of a cell nucleus while preserving all of the cells functions. As early as 1986, Australian biochemist Philip Nagley, at Monash University, succeeded in moving a mitochondrial gene in yeast into the nucleus with no apparent ill effects. While a great deal of work remains to move all thirteen genes, the principle has been demonstrated.

De Grey and his coauthors also propose a set of six other techniques, including the destruction of garbage that builds up inside and outside of cells, and genetically engineering humans to eliminate the key genes that cancer depends on. With a focused researched effort, they believe it would be possible to use these techniques to rejuvenate aged mice and virtually eliminate their aging process within ten years. A result that impressive would have the world clamoring to adapt it to humans. But de Grey's targets are far beyond those of most other researchers in the field of aging. Should he be taken seriously?

De Grey is a controversial figure in aging research, but he's written some insightful and important papers. Some of the coauthors of the paper that presents his anti-aging strategy have impressive credentials themselves. Bruce Ames is one of the most cited biochemists in the world. Andrzej Bartke discovered one of the genes that extends the life span of mice, and also has shown that combining that gene with caloric restriction extends life still further. Judith Campisi heads the department of cellular and molecular biology at the Lawrence Berkeley National Laboratory and has done important work on the links between aging and cancer.

Have these serious scientists in the field of aging gone off the deep end? De Grey's proposals make sense on paper. He makes a plausible argument for a set of deep changes we could make to the human body to eliminate the major causes of aging that we know of. The major case against de Grey's proposal is that it's entirely theoretical at this point. No one has shown that the techniques he proposes lengthen life in animals. In general, biology is an extremely messy, complicated field. What looks right on paper doesn't always work in the lab or in the body.

The best way to find out whether de Grey's proposals make sense is to try them. He's right about that. He's also right that each of the techniques he proposes should be possible in lab animals right now. De Grey is currently attempting to create a research institute to fund and coordinate research along these lines. If he's successful, then in about ten years we'll know whether his ideas actually work.

Whether de Grey succeeds in his more ambitious goal of halting aging or not, there's good reason to hope that we'll see safe and effective age-slowing therapies within the next few decades. Caloric restriction seems likely to work in humans, and researchers are closing in on both genes and drugs that can mimic CR's effects. The sudden flurry of discoveries of new genes that slow aging in lab animals is also just the very beginning. In the next few years, we'll see even more genetic discoveries and further attempts to combine multiple techniques to lengthen life even more. It's not too hard to imagine a strain of mice that live twice as long as normal mice. And with every genetic discovery, there will be additional avenues for biotech and pharmaceutical companies to pursue in creating drugs that slow the aging process.

The leading researchers in extending the lives of animals seem to agree that we'll soon be able to do the same for humans. Prominent and well-respected researchers like Cynthia Kenyon, Tom Perls, Lenny Guarente, George Roth, and Mark Lane have already expressed their beliefs in a very tangible way: by founding companies to bring their discoveries of life extension in animals to human beings. As Kenyon says about extending human life, "The lights are green everywhere you go."

Michael Rose, an evolutionary biologist who created super flies that live twice as long as normal and have greater vigor and endurance, thinks scientists have all but proved that they can manipulate human aging. "Aging is not immutable. It's not God's grace. It is a genetic problem, and you can solve it."

Steve Austad, whose work has been focused on studying the evolution of aging rather than finding specific ways to lengthen life, is also optimistic about bringing the science to humans. "The sorts of therapies that I've been talking about that work in animals [will] actually get extended to humans, and I think that some of them will end up being relevant," he says. "I think it's easily possible that we'll get a few additional decades of human life expectancy." At the same time, he's a bit more conservative on when humans will benefit. In 2002, he said, "I think the time horizon for anti-aging therapies [in humans] is probably something on the order of thirty to sixty years from now."

In contrast, Tom Johnson, the pioneering researcher who first showed that small genetic changes could dramatically lengthen the lives of laboratory animals, has suggested that we'll see effective life-extension drugs on the market by 2020.

Judith Campisi is even more bullish: "We know we can extend the life span of mammals. There is no reason to believe that we couldn't do the same today in humans."

Combine all the evidence and the odds look good that someday soon we'll see a drug or gene therapy that slows the aging process in many different kinds of animals and that seems safe enough for human trials. People who started using such a therapy in their twenties or thirties could have decades added to their lives. The evidence from animal studies suggests that those years would be vital ones: our bodies would look younger, our minds would retain the flexibility of youth, and we could live life to the fullest. Then the field of life extension will hit a whole new era—one in which we'll see the effects that longer lives have on our society.

CHAPTER 6

Methuselah's World

F RESEARCHERS SUCCEED in developing drugs or gene therapies that can slow human aging, who will benefit? We have discussed the price of enhancement and noted that like other technologies, biotech enhancements will likely be expensive when first introduced and then will drop in price over time.

There's another, more specific reason to believe that life-extension techniques will make it out to the masses: the historical record. Throughout most of human history, life expectancy hovered at a very low level—somewhere between eighteen and twenty-five years. Then around 1800, life expectancy began to rise quickly in developed areas: Europe, the United States, and Japan. For the next century, life expectancy rose in these countries while remaining low in the developing world. Eventually, though, sanitation, vaccination, and other public health techniques took hold in developing countries of Asia, South America, and Africa, starting a life-extension revolution there. Ever since, the life-expectancy gap between rich and poor countries has been largely a lag in time.

Life Expectancy in Developed and Developing Countries

FIGURE 6 Developing nations have seen their life expectancies increase later but more quickly than developed countries.

In the 1800s, the rich got a head start in living longer, just as they may with new biotech enhancements. Over the last century, though, the life expectancy of still-developing nations, as defined by the United Nations, has been increasing faster than that of countries the UN defines as already developed. The length of the average life in still-developing nations has gone from twenty-eight to sixty-three years, and is on track to reach seventy-five years in 2050. Meanwhile, life expectancy in developed countries has gone from fifty years in 1900 to seventy-five years in 2000, on its way to eighty-two in 2050 (see figure 6).

Those numbers show that the life-expectancy gap between rich and poor is narrowing. In 1900, a child born in a developing country could expect to live, on average, just over 50 percent as long as one born in a developed country. By 2000, that proportion had climbed to 84 percent. The UN Population Division expects life expectancies in less-developed nations to reach 91 percent of those in developed countries by 2050. Thus, according to the most concrete measure there is, the difference in material well-being between rich and poor has shrunk over the last century, and seems set to shrink further over the coming fifty years (see figure 7).

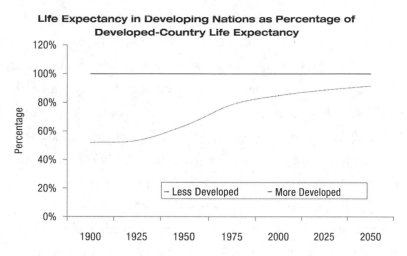

Life Expectancy in Developing Nations as Percentage of Developed-Country Life Expectancy

FIGURE 7 Over the last century, developing countries have closed much of the life-expectancy gap with already-developed nations.

Combined with the evidence for the rapidly dropping price of information goods, the past history of life expectancy suggests that the poor of the world will eventually benefit from biotech enhancements and from age-slowing techniques in particular. The poor won't have access to such techniques as soon as they come on the market, but once they do get access, they may make progress more quickly than the rich and close the gap over time.

Dodging the Demographic Bullet

If age-slowing techniques do become widely available, they could help society avoid the problems of an aging population. In 1900, there were 3 million people over the age of sixty-five in the United States. By 1996, that number had grown more than elevenfold, to 35 million. By 2030, it's expected to double again, to 70 million, and one in five Americans will be over the age of sixty-five. In other countries, the shift will be even more dramatic. In Japan, for instance, the median age of the population is expected to rise to fifty-three by the year 2050; it was twenty-two in 1950.

This demographic shift comes with consequences for everyone. In the United States, as people hit retirement age, they'll leave the workforce, depleting the country's economic output and tax revenues. At the same time, they'll start to collect their Social Security payments and other pensions. As the baby boom generation reaches retirement age, its members will go from being economic contributors to society to an economic drain.

Social Security and other national pension plans are already facing a crisis brought on by this aging population. In 1950, in the United States there were more than sixteen people of working age (twenty to sixty-four) for every person aged sixty-five or older. Today that ratio is closer to three to one. By 2030, it's projected that there will be just two working-age Americans per retirement-age American. If everyone retires at age sixty-five, there will simply not be enough workers generating money to pay for the pensions of the retirees.

What's more, the elderly consume radically more health care than the young. In developed countries, individuals over age sixty-five spend three to five times what people under sixty-five spend on health care. Currently, costs rise exponentially as the end of life nears: 28 percent of Medicare spending goes to patients in the last year of life, and 47 percent of all Medicare spending occurs in the last three years of life (see figure 8, page 117).

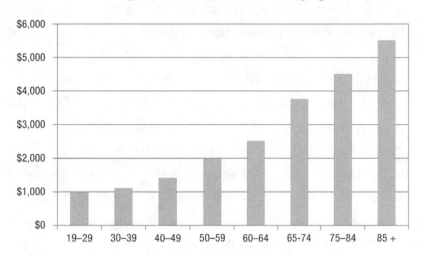

FIGURE 8 Health-care spending in the United States rises dramatically with age as the likelihood of serious ailments increases.

The total cost to society is staggering. In 2001, the over-sixty-five group accounted for almost half of all health-care spending in the United States, despite making up only about 13 percent of the population. More precisely, that much is spent on them. Most health care is paid for by insurance plans or government programs, both of which pass the costs on to younger, healthier individuals. As we invent newer, more expensive life-saving treatments, and as the population of developed countries continues to age, the costs will rise even further.

All told, the Congressional Budget Office projects that Social Security, Medicare, and Medicaid will rise from 7.5 percent of U.S. GDP in 2000 to almost 15 percent in 2030. This means that one out of every seven dollars earned in the United States will go to fund programs for the elderly. The lion's share of this growth isn't in Social Security—it's actually in the cost of Medicare for the old.

The Congressional Budget Office projects that, by 2075, these programs will require one of out every *five* dollars earned in the United States, and their costs will still be rising.

This crisis will hit before life-extension technologies have had much effect on the population. Governments will respond in a variety of ways, including raising the retirement age and reducing benefits. Ultimately, they may go further and link pension payments to ill health or other disability that prevents one from working. After all, when Social Security was first created, its stated purpose was to protect the old from the ravages of "penniless, helpless old age." A sixty-five-year-old who's fit enough to play tennis three times a week and sharp enough to reenroll in college isn't what the founders of Social Security had in mind.

Slowing human aging is one of the few ways out of the Struldbruggian dilemma of an aging population. If the projections of Johnson, Kenyon, de Grey, and other scientists are correct, life-extension technologies will start to make their impact around the time that big changes are due in national pensions. By keeping some of the elderly biologically younger, they'll slow the growth of medical spending on the aging population, combating the largest increase in spending the Congressional Budget Office projects.

Furthermore, life-extended animals display *compressed morbidity:* they frequently remain fit and active until their final days, then die quickly. Sometimes those deaths have no apparent cause. Autopsies of calorically restricted mice find that about one third of the animals have nothing obviously wrong with them at the time of their death. If elderly humans died while still in good health, it would eliminate much of the cost of these final years, and also reduce suffering.

In addition to cost reduction, slowing aging would also inject dollars into the economy. Life-extension techniques would increase the number of elderly who are still mentally and physically sharp enough to work and contribute to the economy. Ironically, as a reward for good genes, good health habits, or good sense in pursuing age-slowing therapies, the fittest and healthiest older people may find themselves required to remain in the workforce. Losing the promise of a long government-funded retirement isn't very pleasant. Even so, most of us would rather be in good health and have to work than be in poor health and get Social Security checks.

Across the developed world, the current national pension systems are in trouble. The Congressional Budget Office projects that without some sort of reform, Social Security, Medicare, and Medicaid combined will consume two thirds of U.S. federal tax dollars by 2030. By 2040, the Social Security trust fund will be empty. To maintain solvency, the United States and other nations will need to either raise taxes dramatically or reduce benefits. Given a healthier workforce, the most obvious step may be to delay the age of retirement. Slowing the rate of aging is one of the few steps that could help delay this demographic crunch.

Life Extension and Population

How will longer lives affect world population? Certainly anything that keeps people alive longer will increase the number alive at any given point. However, the details of population growth and demographics can be rather counterintuitive. Consider that today the countries with the longest life expectancies at birth have populations that are remaining steady or even shrinking. For example, the UN Population Division expects the populations of Japan, Italy, Germany, and Spain to decrease over the next fifty years, despite the fact that Japan has the highest life expectancy of any large country, and the Western European nations are not far behind. At the other extreme, the countries with the fastest-growing populations are those with relatively low life expectancies—India, China, Pakistan, and Nigeria.

Throughout most of human history, birth rates and death rates were both extremely high. In A.D. 1000, birth rates were around 70 births per 1,000 people per year, and death rates were about 69.5 deaths per 1,000 people per year. So each year a town of 1,000 people would on average increase its population by half a person—a 0.05 percent growth rate.

As societies advance, nutrition, sanitation, and medicine all serve to lower the death rate. When births outstrip deaths, population soars, as happened through the twentieth century. But in the last few decades, the birth rate has also dropped sharply, particularly in the rich, developed nations of the world. As countries gain in wealth and education, and especially as women in those countries gain greater rights, resources are increasingly devoted to education and careers rather than to raising a family. This demographic trend is spreading from Europe, Japan, and North America to the rest of the world. Worldwide, the birth rate is now 21 births per year per 1,000 people. The death rate is around 10 deaths per year per 1,000 people. So overall, the world population is growing at roughly 1 percent, slower than at any time in the last few centuries.

Because the birth rate is double the death rate, phenomena that affect fertility can have a far larger impact than similar-size effects on the death rate. For example, between 2000 and 2050, the UN expects around 3.7 billion people on earth to die and 6.6 billion to be born. Cutting the death rate in half would increase population by around 1.9 billion people. Cutting the birth rate in half would reduce the 2050 population by 3.3 billion. Thus the birth rate is a more significant lever on the size of the population.

Projecting population growth into the future is a tricky thing. It depends on just how quickly economic growth improves health care, sanitation, and nutrition. It also depends on how quickly those important birth rates come down. The UN Population Division produces high, medium, and low estimates for future population. Even though birth and death rates differ by just a few percentage points in the different scenarios, over decades it adds up. In 2000, world population stood at 6 billion people. Looking forward, the UN predicts that in 2050 world population will be somewhere between 8 billion and 11 billion, with a "medium" prediction of 8.9 billion.

The figure 8.9 billion people is half again as many as are alive today, yet it represents a growth rate about half that of the last twenty-five years. World population grew by 50 percent between 1975 and 2000. The next 50 percent increase is predicted to take until sometime after 2050, about twice as long. Ultimately the UN's "medium" model projects world population hitting a plateau of 10 billion people in 2100. Of course, projections one hundred years in the future are perilous. They rely on a generally smooth progression of current trends. They don't take into account the possibility of global plagues, world wars, or technologies that radically alter humankind.

Life extension is just one of those technologies, yet its impact on population is surprisingly small. Demographer Jay Olshansky, despite being a pessimist about the prospects for slowing human aging, has shown that extending human life would have an incremental rather than exponential effect on population. "The bottom line is that if we achieved immortality today, the growth rate of the population would be less than what we observed during the post–World War II baby boom," he says. If everyone were made completely immortal today, he calculates, and taking into account declining birth rates, global population would hit about 13 billion in 2100, rather than the 10 billion currently projected.

The immortality Olshansky is talking about isn't achievable. Even if we halted all aging, accidents, homicide, suicide, and infectious diseases would still kill people. Nor is complete halting of the aging process in our sights today. More likely we'll achieve the technology to slow but not halt the rate of aging in the next ten to twenty years. Then it will take additional years for people around the world to gain access to the technology. Even then, the impact on population won't be instant. Those who are already old won't benefit much from techniques that slow their aging rate. Instead, the middle-aged and young will reap most of the benefits. The combination of those factors suggests a gradual impact on worldwide population.

Here's a simple mathematical way to look at this. We can't predict exactly when life-extension techniques will first be available or how quickly they'll spread, so we'll need to make some guesses. Imagine that life-extension techniques first appear on the market in 2015. Let's say that in the year after that (2016), the overall death rate around the world drops by 1 percent as a result. And let's say that it decreases by another 1 percent each year thereafter, so that in 2017 age-slowing techniques have reduced the global death rate by 2 percent and by 2050 they've reduced it by 35 percent. That's quite an optimistic scenario. It implies that by 2050, we'll have managed to increase life expectancy in the developed world to around 120 and in the developing world to around 113—an unprecedented rise in life expectancy, even compared to improvements in the last two centuries.

If this extremely optimistic rise in life expectancy occurred, how large would the impact on global population be? If we simply reduce the number of deaths the UN forecasts each year by the percentages just given (1 percent in 2016 climbing to 35 percent in 2050), we get a total population in 2050 of 9.4 billion people, instead of the 8.9 billion the UN projects.

Five hundred million additional people are nothing to sneeze at, but it's less than a 6 percent increase in the projected world population for 2050. To put this in context, that's less than the percentage change in world population between 1970 and 1973—a noticeable change, to be sure, but not a catastrophic one.

It's also worth noting that some demographers believe that birth rates in the developing world will drop faster than the UN's middle scenario. Just a 5 percent lower birth rate yields a scenario where global population tops out at around 8 billion in 2050, and may by 2100 fall back to below the 2000 level of 6 billion. Achieving this is entirely plausible if the right resources are applied. A plateau and eventual reduction in global population depends on the spread of wealth, education, and freedom in the developing world. For example, the UN expects the population of Europe (not counting immigration) to shrink by about 0.4 percent per year between now and 2050. The overall population of the developed world (again not counting immigration) is expected to remain fairly constant. If the developing world can attain the present level of affluence of the developed world by 2050, we'll see overall world population level out and begin to drop.

Enhancement and Social Change

Health care, retirement, and population are all fairly clear-cut issues in which trends lend themselves to mathematical analysis. Yet the prospect of conquering aging raises other, "softer" issues. In particular, how will longer lives affect social change and progress? Many human organizations are hierarchical: people struggle to get to the top, where they can wield real power, and once they get there, they don't want to let go. Business executives, incumbent politicians, and university department heads are just a few examples of this.

If human lives are lengthened, we can expect the people at the top to try to hold on to their positions, and in so doing they'll be robbing younger generations of opportunity. They may also serve to forestall innovation. In some academic fields, there's a saying that progress occurs one death at a time, as old stalwarts die and a younger generation can finally get its ideas out. With older people clinging to life and power, this may not be possible. Indeed, we may see more intergenerational conflict between the very old and traditional on one hand and new generations willing to adopt more radical technologies and ideas on the other.

Social, political, and academic stagnation and intergenerational conflict are plausible risks. Yet there are also powerful forces working against stagnation of the old. Age-retarding technologies don't just extend life, after all. They keep our brains and bodies younger. Younger brains learn faster and adapt to new conditions more easily. A sixty-five-year-old in the middle of this century may have a brain as flexible as that of a twenty-five-year old, yet bolstered by all the accumulated experience of sixty-five years of life.

A second consideration is our increasing ability to sculpt our own minds to boost learning rate or memory, or even to alter facets of personality. Competitive pressures will fuel the adoption of mind-enhancing and sculpting technologies among the life-extended. It's one thing for a manager or business leader to become stuck in his or her ways. It's quite another thing for a whole company or university to become stagnant. When that happens, other more flexible competitors will swoop in to take their slice of the pie. At an organizational level, there's a kind of natural selection. Organizations that are slow to change will die out. The nimble ones will probably be those with nimble people in them.

An older population is likely to trigger other changes in society. For example, older people are more likely to vote. In the U.S. election of 2000, only 34 percent of citizens aged eighteen to twenty-four voted, whereas 73 percent of citizens aged sixty-five to seventy-four voted. In addition, older people are far less likely to commit violent crimes. The large majority of violent crimes are committed by young men aged eighteen to twenty-four. Polls have also found that the elderly are also less likely to support military action as a way to solve international conflicts than the general population. Combined, these suggest that an older population is both more engaged in civic life and more restrained in the use of violence. How that will impact society in the long term is difficult to say, but as the population ages we will likely find out, with or without life extension.

In the end, the practical benefits of life extension seem to outweigh the risks. We know that slowing aging would be the most effective way to combat a wide variety of diseases, that it would soften the economic blow of the graying of society, and that it would have a surprisingly small impact on global population. Given those more concrete data points, it seems most reasonable to take a wait-and-see attitude toward the secondary effects that life extension will have on society. Undoubtedly, longer lives will cause unforeseen problems, and as a society we'll have to come together to solve them, just as we've solved the problems that have come with previous social goods.

Ultimately, longer lives can't be considered in a vacuum. Life extension is just one of several transformative technologies humanity is developing. The other technologies discussed in this book will give men and women greater ability to transcend their present limitations, to reinvent themselves, to learn and improve beyond the plateaus we now encounter. As we learn to redesign ourselves, we'll reach states of mental and emotional capacity for growth that simply can't be satisfied in one human lifetime.

When we contemplate the three years that a mouse may live, we don't mourn its short time on this earth. In three years, a mouse lives and learns as much as it's able, and more years wouldn't add meaning or quality to its life. Today a human life span may provide enough years for a man or woman to learn and grow as much as we're able. But in the decades to come, we'll increase our capacity to learn, grow, and change over time. Eventually one hundred years may seem like a brief adolescence—an early phase of life in which we learn many things but only barely reach maturity.

Today we mourn those who die in adolescence, because they have so much left to see and do in their lives. As we push back the limits on our minds and bodies, as we increase our own ability to grow and learn, we'll feel the same way about those who die at eighty or one hundred. With the power to transform ourselves, our futures become less limited by our starting point, and more by the lengths of our lives. By extending those lives, we'll give ourselves, as individuals and as a society, room to grow and blossom in ways that we have yet to imagine.

CHAPTER 7

A Child
of One's
Own

IN THE LATE 1960s, Lesley and John Brown, a young couple from Bristol, England, decided to try to have a baby. Nine years later, Lesley still had not conceived. As it turns out, Lesley's Fallopian tubes were blocked, so her eggs couldn't pass down from her ovaries to meet her husband's sperm and fuse to form a zygote, a single-celled embryo. Over those nine years, John and Lesley went from doctor to doctor, but none was able to help.

In 1976, the Browns were referred to Patrick Steptoe, a gynecologist at Oldham General Hospital, who was ten years into his own quest into new ways to help infertile couples through a technique called in vitro fertilization (IVF). With his colleague Robert Edwards at Cambridge University, he'd been able to combine sperm and egg in a lab dish and produce early embryos. They'd implanted these embryos into their prospective mothers, but as yet none of the pregnancies had lasted longer than a few weeks.

Still, Steptoe and Edwards were optimistic. In their previous attempts, they'd implanted the embryo into the prospective mother's body five days after conception, when it had reached a size of sixty-four cells. They theorized that by implanting the embryo into the mother's body earlier, when it was only eight cells, they might improve the odds of success.

On November 10, 1977, Steptoe used a probe called a laparoscope to remove an egg from one of Lesley Brown's ovaries. Edwards mixed the egg with some of John Brown's sperm. One of those sperm penetrated the wall of Lesley's egg. Conception had occurred.

Two days later, Steptoe carefully inserted the eight-cell embryo into Lesley's uterus. This was a tricky point in the process. In natural conception, more than half of all embryos at this stage are flushed from the body without ever implanting in the uterine wall. But luck won out, and the embryo implanted in Lesley's body. She was pregnant.

Eight and a half months later, on July 25, 1978, Lesley Brown gave birth to Louise Joy Brown, a healthy, blue-eyed, blond baby girl, and the world's first "test-tube baby" had been born. Through IVF, science had given us the ability to create—or at least conceive—life in the laboratory.

The initial public response to IVF wasn't pretty. Jeremy Rifkin and other critics of biotechnology (including Leon Kass, now chairman of the President's Council on Bioethics) criticized the procedure. Edwards and Steptoe were accused of playing God. After their second successful IVF delivery, public protest forced Steptoe and Edwards to halt their work for two years.

After the initial shock wore off, however, the technique rapidly gained in popularity. Since 1978, more than a million babies conceived by IVF have been born. More than 100,000 more are born each year. One out of every hundred births in the United States is conceived through IVF. In the United Kingdom, France, and other parts of Europe, where the cost of the procedure is more frequently subsidized, the number is one in every fifty births. In Australia, one in every *twenty* births is conceived through IVF.

The process is still essentially the same as when John and Lesley Brown went through it. Doctors take several eggs from the prospective mother and sperm from the prospective father and mix them in a lab dish. If all goes well, a handful of the eggs will be fertilized by sperm and become embryos. Forty-eight to seventy-two hours later, two or three of the healthiest-looking embryos, each of which has four to eight cells, are implanted back into the mother to begin her pregnancy.

IVF may sound simple, but it's actually a rather challenging procedure. First, there is the cost. A single IVF cycle can cost anywhere from $5,000 to $15,000. Then there is the trauma to the mother. A woman going through an IVF cycle faces a daunting array of injections, tests, and procedures to determine her health, promote production of several eggs, extract those eggs, and implant the fertilized embryo(s) back into her uterus. All in all it's a physically, emotionally, and financially grueling process stretched out over at least two weeks. And there are no guarantees that a woman will get pregnant in a single cycle. Often the embryos fail to implant in her uterus, and the entire process must be repeated.

Despite this, IVF has become mainstream, and the reason is very simple. Worldwide, as many as one in every six couples is infertile. Without IVF and other assisted-reproduction techniques, they cannot have biological children, and the urge to have offspring is one of the most powerful instincts that we carry. It's embedded deep in our genes and in the genes of every other species that has ever lived on this planet.

Faced with the choice of having no biological children or going through an expensive, painful, and uncertain procedure in order to have them, more and more couples are choosing the latter. This basic desire to reproduce fuels IVF, and is now fueling other techniques aimed at bringing healthy babies into this world.

Susan and Chris Dunthorne had a different problem than Louise Joy Brown's parents. They weren't infertile—the birth of their son Joshua proved that. But Joshua died just four months later, a victim of cystic fibrosis.

Cystic fibrosis is caused by a mutation in a gene on the seventh chromosome called CFTR. The disease is recessive, meaning that it only manifests in children who inherit two defective copies of the CFTR genes. About one in every twenty Americans carries one broken copy of this gene. Thirty thousand people in the United States carry two broken copies and suffer from the disease.

The Dunthornes were heartbroken at the death of their son. They desperately wanted another child, but couldn't in good conscience take the risk of inflicting cystic fibrosis on another boy or girl. Either watching another child die or watching him or her grow up afflicted with the disease promised to be too painful. As a result, they feared they'd never be able to have a child of their own.

That was before the Dunthornes learned about preimplantation genetic diagnosis, or PGD. PGD is a step that has been added to in vitro fertilization. It was first developed to increase the success rate of IVF, by helping doctors pick the healthiest embryos— the ones most likely to implant in their mother's uterus and lead to a successful pregnancy.

In PGD, mother and doctor go through the normal steps of IVF. After embryos are conceived outside the body, they're allowed to develop for two or three days. At this point, when the embryos have reached the eight-cell stage, doctors remove one or two of these cells from each embryo and test them for the presence of healthy chromosomes. (The other six or seven cells go on just fine without the ones that have been removed; at this early stage, all the cells in the embryo are identical, and each one is capable of producing any of the hundreds of kinds of cells found in a fetus.)

The tests run as part of PGD can spot the difference between healthy embryos that are likely to lead to pregnancy and chromosomally malformed embryos that are more likely to be flushed out of the body or lead to a miscarriage. By spotting major problems in chromosome structure, PGD has already proven successful at increasing the success rate of IVF, especially among couples with a history of many IVF failures.

PGD can also be used to spot genetic diseases. Once doctors have genetic material from the embryo, they can test it for the presence of any gene that has been identified, including defective CFTR. In the last decade, parents and doctors have used PGD to test for Down syndrome, sickle-cell anemia, muscular dystrophy, Huntington's disease, and other genetic disorders. All in all, several thousand PGD procedures have now been performed since the technique was first reported in 1989.

PGD offered the Dunthornes new hope. Their physicians at Hammersmith Hospital in London withdrew fifteen eggs from Susan's ovaries and mixed them with Chris's sperm. Six of the eggs were fertilized and developed into embryos. The doctors removed one cell from each of the embryos for testing. The first round of tests failed to provide the data needed, and hope almost collapsed. But the second round succeeded, showing that two of the embryos were healthy and free from the CFTR mutations that would lead to cystic fibrosis. Both embryos were implanted in Susan's uterus, and one of them attached itself within her. Ten days later they got the news—Susan was pregnant. Nine months later their baby son, Ethan, was born, free of cystic fibrosis.

This is just the beginning for PGD. The cutting edge has moved from using the technique to spot defective genes to looking for genes that increase the risk of contracting a disease later in life. For example, PGD has been used to spot dangerous mutations of the BRCA1 gene, which increase the likelihood of cancer. It's been proposed to spot dangerous variants of the ApoE gene that increase the likelihood of Alzheimer's disease. Several PGD centers in the United States will now attempt PGD for any condition for which there is a genetic test—a list that is growing day by day.

PGD does have major limitations, though. You have thirty thousand genes. You differ from other humans in as many as 3 million ways. Yet IVF procedures typically produce only a handful of embryos. None of those embryos will ever be perfect. Indeed, the more we know about genetics, the more obvious the imperfections will be.

In Chris and Susan's case, a different outcome of the genetic lottery might not have worked out so well. Of the six embryos they conceived, only two were healthy and free of cystic fibrosis. What if all of the embryos they'd conceived had carried the CFTR genes? Or if one of the CFTR-free embryos had carried a gene with a high risk factor for Alzheimer's disease, or Parkinson's? If every embryo a couple conceives carries a high risk factor for *some* disease, what can they do?

There are in fact couples for whom this will always be the case. For example, there are three variants, or *alleles,* of the ApoE gene—e2, e3, and e4. People who carry one copy of the e4 allele have an elevated risk of both Alzheimer's disease and high cholesterol. People who carry two copies of the e4 allele have a 60 percent chance of contracting Alzheimer's disease. If an individual carries two copies of the e4 allele of ApoE, then all of his or her children will be guaranteed to inherit at least one copy. If *both* mother and father carry two copies of e4, then all of his or her children will also have two copies, and a 60 percent chance of having Alzheimer's.

For parents like these, no amount of genetic screening can remove the risk factors from their offspring. No matter how many embryos they conceive, their children's possible genes can only be chosen from what their parents carry. Today such parents have two options. They can have a child and hope that medicine solves the problem by the time he or she grows up. Or they can elect to not have a biological child and either adopt or not raise any children.

Eventually, though, some couples won't be satisfied with these options. They'll ask their doctor, "Isn't there something more we can do? Can't we just tweak this one little gene? Can't we change that dangerous e4 into a healthy e2?" Thousands of prospective parents will ask questions like this, to which their physicians will reply "no." Until one day some sufficiently motivated parents find a doctor who answers, "Yes. We can tweak that one little gene."

To date, no one has altered the genes of a human embryo. The closest researchers have come to this is a rhesus monkey named ANDi. ANDi was the first genetically engineered primate brought into this world. Her creators took a rhesus monkey egg and inserted a jellyfish gene, one that makes jellyfish glow green. The gene is in every cell in ANDi's body. By inserting a foreign gene into an embryo of such a close relative of ours, scientists have demonstrated that genetically engineering human embryos is nearly feasible.

The technique for genetically engineering an embryo is similar to the gene therapy discussed earlier: scientists use a gene vector to deliver a new gene to the embryo in essentially the same way they would deliver a new gene to any other cell. Scientifically and technologically, the two techniques are almost the same.

There are important differences in the results produced when a gene is inserted at such an early stage, though. To illustrate these differences, let's compare ANDi, the genetically engineered monkey we just talked about, and Ashanti DeSilva, the first human to receive gene therapy.

ANDi went through *germline genetic engineering*. She had a new gene inserted into her when she was just a single fertilized cell. That gene integrated itself into the DNA of that single cell, and from then on, every time that cell and its daughter cells divided and made new copies, the new DNA was copied too. Consequently, every cell in ANDi's body carries the jellyfish protein gene, and if ANDi has children, she has a 50 percent chance of passing that gene on to them as well. Her germline—the genetic information she can convey to future generations—has been engineered, hence the term germline genetic engineering.

Ashanti, on the other hand, underwent *somatic gene therapy*. She was four years old when it happened. By this time, Ashanti's body was made up of trillions of cells. Delivering the new gene to every one of her cells was effectively impossible, so researchers focused on just the most critical cells for her disease—blood cells. Somatic cells are cells that, like blood cells, aren't involved in reproduction, so if Ashanti has children, they'll inherit their genes from among the pool that Ashanti was born with, not the version that French Anderson and his colleagues injected into her blood.

Altering the genes of a person before he or she is born is in some ways actually easier than doing so after birth. It's easier to get a gene into every one of a small number of cells than into a large number. At early stages of development, there's no immune system in the embryo to respond in potentially dangerous ways to the gene vector. So as medical motivations drive improvements and innovations in gene therapy, they will also tend to drive germline genetic engineering. As researchers increase the safety and effectiveness of somatic gene therapy through new and better gene vectors, indirectly they'll be making genetic engineering safer and more effective as well.

There are already proposals for procedures that would blur the line between somatic gene therapy and germline engineering. In 1998, French Anderson proposed in utero gene therapy. His idea was to perform gene therapy on a few-weeks- to few-months-old fetus suffering from a genetic ailment, while the fetus was still in its mother's womb. The Recombinant Advisory Commission, the watchdog that looks over all uses of genetic technology in the United States, turned down the request. The commission was concerned that if gene therapy were performed during prenatal development, the inserted genes might find their way to the child's sex organs and then be passed down to future generations.

Despite this, there are medical reasons to think about the feasibility of in utero gene therapy. Most children are still conceived the old-fashioned way, through sex rather than IVF. In these children, there's no chance to do genetic screening or alteration before the embryo implants in the womb. But around 3 percent of all pregnancies result in a child with a birth defect of one sort or other, the most common being congenital heart defects, to which several genes have been found to contribute.

Increasingly, concern around the health of unborn children has led to prenatal genetic testing. In the United States around 13 percent of all expectant mothers go through some form of prenatal genetic test during pregnancy. Amniocentesis, the best known of these procedures, is now recommended for any expectant mother over the age of thirty-five, or for those with a history of Down syndrome or other genetic disorder in their family.

In January 2004, a team of researchers and physicians from the University of California at San Francisco and Washington University published a paper in the British journal *The Lancet* recommending that all mothers undergo amniocentesis or another prenatal genetic test. The team, led by Miriam Kuppermann of UCSF, weighed the pros and cons in economic terms. Amniocentesis has a cost, and it also carries with it a 0.25 to 0.50 percent risk of miscarriage. On the other hand, it can help identify Down syndrome and other problems early in a pregnancy. The total savings to society by early detection of fetal illness, Kuppermann showed, outweighed the cost and risk of the procedure.

What's more, newer tests like CVS (chorionic villi sampling) are less expensive and can be performed earlier in the pregnancy. These tests can pinpoint quite a few genetic diseases, including some that do irreversible harm before birth. For example, Tay-Sachs disease does permanent damage to the brain during gestation. In the near future, prenatal tests will be able to detect Tay-Sachs while an embryo is just a few weeks old.

By performing gene therapy early in a pregnancy, doctors could potentially eliminate Tay-Sachs, cystic fibrosis, congenital heart problems, or other genetic diseases before they have a chance to harm the child. And because the embryo is still rapidly developing, any genes inserted during the pregnancy will be spread through more of the body. In effect, the earlier gene therapy is performed, the easier it is.

For now, the Recombinant Advisory Commission's ruling effectively forbids in utero gene therapy. Yet as genetic testing becomes faster, easier, and more common, more prospective parents will find out that their unborn children bear some genetic disease. Those parents will be looking for ways to cure those diseases as soon as possible.

In utero gene therapy offers such a route. And despite widespread unease about genetic technologies, most Americans seem willing to employ them for medical uses. In a 2002 survey of more than twelve hundred people, the Johns Hopkins University Genetics and Public Policy Center found that 73 percent approved of the use of PGD to preempt genetic disease. The same poll found that 59 percent of Americans approved of the use of genetic engineering to eliminate disease genes from the unborn.

That's a staggering number. It suggests that of the 4 million children who will be born in the United States this year, around 2.4 million of them will be born to parents who approve of the use of genetic engineering to cure genetic diseases. Of the 120,000 American children who will be born with some sort of birth defect this year, more than 70,000 of them have parents who approve of the use of genetic engineering to correct those defects while the child is still in the womb.

In vitro fertilization, preimplantation genetic diagnosis, in utero gene therapy, and genetic engineering have all been developed in service to one thing—the human desire to have and raise a healthy child. The statistics we just discussed are a testament to how strong that desire is. IVF has already helped more than 3 million couples conceive. PGD has helped bring several thousand children into this world free of crippling genetic defects. In utero gene therapy and genetic engineering will eventually help other couples and unborn children.

As with other medical technologies, society stands to benefit from these reproductive-health techniques through a reduction in health-care costs of children born with diseases. PGD is expensive, but far less so than chemotherapy or a lifetime of care for a victim of Down syndrome. Prevention is easier than cure. If we can use technology to reduce the incidence of disease-causing or disease-promoting genes, we all stand to benefit, and none more than the children born from these techniques themselves.

That hasn't stopped critics from attacking these techniques. The most commonly voiced objection is that meddling in reproduction is unnatural or constitutes an attempt to play God. Yet as noted, that argument ignores much of human history and many facets of modern society. We've already made major changes in reproduction since our hunter-gatherer days. We use contraceptives to decouple sex from fertility. We use ultrasounds and prenatal tests to learn about our children before they're born. Our doctors perform cesarean sections to deliver difficult pregnancies. We use incubators and medicines to save the lives of sick or premature babies.

In short, we already intervene at every point we can to stack the deck in favor of having a healthy child. These new technologies are just new tools to accomplish a goal as old as our species—to bring our children safely into this world.

CHAPTER 8

A Child of Choice

CRAIG VENTER HAS been called many things—arrogant, brash, disdainful, egomaniacal, even "Darth Venter." He's seldom been called humble. The tall, intense, balding scientist has never been one to back away from a fight. It takes confidence to believe that one can challenge and beat the most ambitious scientific project of the last half century. In 1997, that's what the maverick biologist did when he founded Celera Genomics, a company devoted to sequencing the human genome in just three years. At that point, the publicly funded Human Genome Project (HGP), an international collaborative research program run by the National Human Genome Research Institute (NHGRI) of the National Institutes of Health, was about 10 percent done with its own sequencing of human DNA. The HGP was funded to the tune of more than $10 billion and was on track to finish its work in another five years. Yet with only $300 million of corporate backing and a late start, Venter set out to beat them.

Venter succeeded in sequencing the human genome in those

three short years, and his challenge spurred the public HGP forward, so that the two groups simultaneously released their results in 2000, two years ahead of the public project's original schedule.

Journalist James Shreeve does a masterful job of telling the story of the personalities in the race to sequence the genome in *The Genome War.* Yet underlying those personal stories is a story of technology and how its progress will affect our world.

The key to Celera's success was a technique called shotgun sequencing. In normal gene sequencing, the long strands of DNA that make up a genome are first mapped out to identify key landmarks. Then the DNA is copied several times and divided into shorter overlapping segments. Those segments can then be read to determine their exact sequence. Researchers refer back to the map to determine where this particular sequence fits in the grand scheme. If you imagine the genome as a giant jigsaw puzzle, it's as if the researchers assemble the entire puzzle, then take each piece, carefully note where it is, and sequence it. By remembering where in the puzzle it belongs, they can then assemble the entire sequence.

Shotgun sequencing isn't so careful. It chops DNA up into short overlapping strands, but doesn't take any note of a map or where those strands fit. Instead, after sequencing each strand, a computer uses the overlaps between the strands to fit them all back together. The researchers never put the jigsaw puzzle together in the first place, which saves them time and money. The Human Genome Project avoided shotgun sequencing as too radical and too dependent on computers, but Venter made a bet on it, and the bet paid off—Celera was able to sequence the human genome at 3 to 5 percent of the cost of the public project, and substantially faster as well.

Shotgun sequencing was not the first speed and cost improvement in sequencing technology, nor was it the last. According to the NHGRI, the cost of sequencing a single base pair, or single "letter," of DNA, dropped by a factor of more than one hundred between 1990 and 2002, from about $10 per base pair to about 9 cents.

NHGRI's report on the topic states, "Starting in 1990 . . . the reduction in sequencing costs has followed a straight line on a log curve." A pattern that follows a straight line on a log curve is one that's growing (or shrinking) exponentially. It means that the amount of DNA you can sequence for a given price seems to be going up by about 50 percent each year, or a little more than doubling every two years.

Additional data going back all the way to the sequencing of RNA in 1965 seem to confirm the general trend, and the experts believe that it will continue. The NHGRI report states that "a two-log [factor of 100] decrease in cost [of sequencing] in the next five years" is possible and that "there is no physical reason that we can't achieve a $1,000 genome." At the current rate of progress, we'll be able to sequence the entire genome of a human being from scratch for $1,000 in the year 2035. Six years later, in 2041, we'll be able to sequence an entire human genome for just $100.

This trend has caught the attention of commentators, scientists, and entrepreneurs alike. Computer scientist Raymond Kurzweil and zoologist Richard Dawkins have both written about this exponential improvement, and connected it to the other famous exponential technology improvement of our time— Moore's Law. Moore's Law, coined by Gordon Moore, the founder of Intel, is the observation that the number of circuits that can be placed on a computer chip, and thus the chip's processing power, is growing exponentially, doubling every eighteen months or two years, depending on whom you ask. It's the driving force behind the ever increasing power and dropping cost of computers.

As it turns out, Moore's Law isn't just an analogy for our rapidly increasing gene-sequencing power, it's also the *basis* of much of that power. The newest techniques for reading genes are actually built with the same technologies used to build the CPU and memory chips inside your computer.

Take gene chips as an example. Gene chips are manufactured using processes taken from the computer industry. Each gene chip is a grid with anywhere from a few thousand to more than a million strands of DNA anchored to its surface. Each tiny square on the grid carries a different DNA sequence. Those DNA strands act like genetic Velcro. When they come in contact with a matching strand, they adhere to it, forming the famous double helix. This allows researchers to identify the genes in a sample by seeing which points on the chip they stick to.

The power of a gene chip is directly related to how fine the grid is. The more distinct DNA sequences you can fit on a chip, the more genes you can search for in a blood sample. As the computer industry comes up with new methods for increasing the density of electronic chips, those techniques become available for gene chips as well. In a very real sense, the power of gene chips is pegged to Moore's Law and the computer industry.

The same is true of another technique, linear DNA analysis. Inside every cell of your body, your DNA is packed into a tight tangle. Every time one of your cell divides, your entire genome must be stretched out so the DNA can be "read" and copied by enzymes. The process takes about fifteen minutes. Linear DNA analysis works similarly. Instead of trying to read short segments of DNA like a gene chip or gene sequencer, linear DNA analyzers stretch out whole chromosomes and run them through a computer chip where a laser can scan them. Linear DNA analyzers, like gene chips, are built with techniques borrowed from the computer industry.

Gene chips and linear DNA analyzers can also cheat a bit. They don't actually sequence your DNA from scratch. Instead, they rely on the fact that all humans have very similar DNA. Looking at every bit of your DNA would be a daunting task, but looking for the smaller number of places where your DNA might differ from the rest of humanity is more manageable.

For all that we've talked about variation in human genes, the differences are actually very small. Your genome is spelled out in pairs of molecules called base pairs, which come in four types: guanine, cytosine, thymine, and adenine—commonly abbreviated as G, C, T, and A. You have about 3 billion of these genetic letters, yet only about one in one thousand base pairs is actually different from person to person. These 3 million places where humans differ are called single nucleotide polymorphisms, or SNPs. (*Nucleotide* is another name for base pair, and *polymorphism* means a place where the nucleotides come in multiple forms or versions.)

The fact that the DNA each individual carries is unique in only 3 million places out of 3 billion makes the task of genetic screening much simpler. Instead of having to read every letter one at a time, researchers can look for whole stretches that are known to come in just one of a few versions. For example, one stretch of DNA may appear in humans in only two versions, one that reads TTGCACG and another that reads TTGCGCG.

By the time you read this book, the leading gene-chip company Affymetrix will be selling gene chips that detect at least one hundred thousand of these SNPs in the span of a few hours. That's still a small chunk of the 3 million SNPs that humans possess, but it's a huge step up in the number of genes that can be identified in a single test. Analyzing 100,000 SNPs by sequencing your genes from scratch would cost several million dollars today, while one of Affymetrix's chips costs just a few thousand dollars.

U.S. Genomics, a start-up company leading the charge toward linear DNA analysis, claims that its system can already scan 30 million base pairs a second for the single nucleotide polymorphisms that distinguish individuals. The company estimates that in the next ten years, it will be able to map an entire genome in a day.

Affymetrix and U.S. Genomics are just two companies in the race to provide faster, cheaper genetic screening technologies. If they fail, their competitors may yet succeed. Driven by the huge potential market, dozens of companies are working to deliver better ways to read the genome. Collectively they're fueling an exponential increase in gene-sequencing and gene-mapping power.

If sequencing power for a given price really is following this Moore's Law–like progression, then the future of sequencing will look something like the future of computing. Today we have cell phones, digital watches, and calculators with more computing power than the supercomputers of a few decades ago. In a few decades, we may have portable lab sequencers with more sequencing power than the entire Human Genome Project.

Cheap gene sequencing would have two immediate consequences. First, the cheaper and easier it is to sequence a genome, the easier it is to determine which genes influence which mental and physical traits. Researchers try to do that today with population genetic studies. For instance, they take a number of people with a certain rare disease and try to determine what genes they have in common with each other but not with the general population. Currently those studies are long and arduous. The tools researchers have for looking at genetic samples are fairly crude and slow. It can take years of research to identify a single gene.

In contrast, within the next decade or two, some of the techniques we just talked about may have advanced to the point where they could be used to map a person's entire genome in a matter of hours. That gain in speed will accelerate research into the functions of our genes and will give us a more comprehensive view of the genes involved in heart disease, cancer, and other ailments.

Inexpensive gene sequencing won't stop at decoding the genes responsible for diseases, though. The technology will give researchers the tools they need to identify genes that influence physical traits like height, facial structure, musculature, and eye, skin, and hair color. The same tools will help identify genes involved in mental traits too—those that impact our intelligence, mood, and personality.

That leads us to the second consequence of cheap gene sequencing: it will make genetic testing of the unborn easier, faster, and more accurate. Where today prenatal tests can identify a few dozen different disease-causing genes, in the near future parents will be able to identify every one of the thirty thousand or so genes their unborn child carries. Combined with knowledge of what these genes do, this information will give parents an idea of a prospective child's appearance, intelligence, and personality.

The Power to Choose

With knowledge comes power. A couple undergoing in vitro fertilization could use preimplantation genetic diagnosis to completely screen the genes of the embryos they've created and select a few for implantation on the basis of the traits they find.

Princeton's Lee Silver, in his 1997 book *Remaking Eden,* paints an evocative picture of how this might work. A couple goes into an infertility lab and goes through IVF. The IVF procedure produces several embryos, each of which has its genes analyzed by PGD. Now the couple sits in front of a computer screen. Displayed on the screen are genetic "profiles" of the embryos they've created and computer-generated pictures showing the children and adults those embryos are likely to grow up to be.

The parents click on one of the profiles and see the face of a potential child at age sixteen. After all, the shape of a nose, the color of eyes, the tone of skin, and much more are encoded in our genes. Each profile also shows any severe genetic diseases the embryo carries, how high the embryo's risk is of developing more complex diseases like heart disease and cancer, and even an assessment of the child's likely personality. *Click*—a lovely green-eyed, brown-haired girl, but with a high risk of heart disease and only average intelligence . . . *Click*—a pudgy, probably shy boy likely to have a high IQ . . . *Click*—a tall, athletic girl with perfect pitch but a tendency toward manic depression. *Click*—another possible child . . . *click*—yet another . . . Several embryos and several possible children.

Yet PGD is inherently limited. Parents can only choose from the set of embryos created in a single round of IVF. This can be anywhere from one to twenty, but seldom more. Each embryo will have some pluses and minuses. One embryo might carry genes for great looks and high potential intelligence, but also a propensity for anxiety. Another might carry a propensity for a sunny disposition and perfect pitch but a propensity for obesity.

Doubtless, some parents will draw the line at merely peeking at their future children's genes. Others, those going through IVF, will need to choose which embryos to implant anyway. At least some of them will be interested in making that decision based on these other factors that genetic tests reveal, but will be content in simply selecting one of the available embryos without any attempt to alter its genes.

Yet, as we discussed last chapter, there will be scenarios in which parents will have good medical reasons to alter the genes of their unborn children. Once those medical uses of genetic engineering or in utero gene therapy are accepted, it may be a short step to altering genes for more cosmetic or enhancement reasons.

Westerners are not very friendly to the idea of using genetic technology to enhance the unborn. Only 22 percent of Americans approve of using PGD to select desirable features, while 73 percent favor using PGD to screen out disease. Only 20 percent of Americans approve of using genetic engineering to create desirable features in the unborn, while 59 percent approve of using genetic engineering to remove genetic diseases.

As we saw a few chapters ago, support for genetic technologies is much higher in other parts of the world, especially Asia, where majorities in several countries approve of the use of genetic engineering to select traits of children. It's possible that attitudes will change in Western nations as well. According to the Johns Hopkins University Genetics and Public Policy Center, the percentage of Americans who approved of the use of genetic engineering to create desired traits in children went from 10 percent in 1994 to 20 percent in 2002. What's more, the attitudes are related to age: only 14 percent of those in the fifty-and-older age group are highly supportive of genetic technologies (as measured by the

Johns Hopkins study) as opposed to 27 percent of those in the eighteen- to twenty-nine age group.

There's also a large gray area between the widely supported medical uses of genetic technologies and the less-supported enhancement uses. Preventing Down syndrome is clearly a medical use. Screening out or using genetic engineering to remove a gene that increases the likelihood of some disease is also medical. But what about selecting for or engineering in a gene that dramatically reduces the risk of heart disease or cancer to far below the population norm? For example, the ApoE e4 allele dramatically increases the risk of both Alzheimer's disease and high cholesterol, whereas the ApoE e2 allele reduces risk below the population norm. Would removing the e4 allele and splicing in the e2 allele be an enhancement or a medical use? Opinions could differ.

Or consider the CCR-5 gene. About 1 percent of the population carries a variant of this gene that makes them immune to most strains of HIV. While altering a child's genome to carry this variant would have no known effect other than resistance to a disease (AIDS), it could still be viewed as an enhancement, as it conveys a protective trait that few members of the population have today.

As a final case, consider obesity, which is a risk factor for heart disease, diabetes, and cancer. At least forty different mutations of six human genes are thought to increase the likelihood of obesity, and reducing the risk of the condition clearly has medical value. Just as clearly it also has a cosmetic value. On medical grounds society might allow genetic engineering to reduce the risk of obesity, yet there would be no way to know whether parents are choosing a gene because they want their child to be healthy or because they want their child to be attractive. No matter what the motive, engineering a child's future weight is certainly cosmetic.

If it's allowed, then people will start to wonder why they can't genetically engineer height, or hair color, or skin tone.

There are dozens of other examples like these, and science will only discover more as time goes on. These gray areas form the slippery slope from therapy to enhancement. Once you accept the idea of genetically altering an embryo to remove a gene that might cause cancer, it's not a far step to adding a gene that reduces the risk of cancer. Once you accept adding a gene that reduces the risk of a disease, a gene that increases youthful life span and thus reduces the risk of many diseases seems more reasonable. Once society is comfortable with splicing in genes to reduce a child's risk of obesity and thus stave off heart disease and diabetes, choosing genes that promote good looks or intelligence doesn't seem so shocking.

History suggests that in general new biological technologies are feared at first, and then accepted as they prove their worth. IVF, organ transplants, vaccination, and other technologies were met with hostility when they first appeared but are now considered routine. Genetic enhancements could follow the same pattern. In Asia, a majority of the population already supports the use of genetic technologies to enhance children. In the West, it seems likely that the strong support for medical uses of genetic technology will gradually shift to bolster the small but already growing support for genetic enhancements.

Cloning

The most widely feared and denounced genetic technology is reproductive cloning—the production of a human baby all of whose genes have been copied from another person. Researchers have cloned sheep, cows, and monkeys, and dubious claims have been made that humans have been cloned. Whether that's happened yet or not, the fundamental technology exists, and the claims have roused public fears.

Cloning has been almost uniformly denounced as morally repugnant. Leon Kass, chairman of the President's Council on Bioethics, describes human cloning as a threat to "the dignity of human procreation" and a "first step toward a eugenic world in which children become objects of manipulation and products of will." Virtually every political leader in every country is publicly opposed to cloned babies. At the United Nations, more than one hundred countries voted in support of a ban on reproductive cloning, though the ban has not yet been ratified, due to disagreement over whether therapeutic cloning, which produces medically useful stem cells but no children, should also be banned.

There are certain cases where reproductive cloning is the best or only way for parents to have a biologically related child. For example, consider a couple in which neither the man's sperm nor the woman's eggs are viable. Neither IVF nor egg donors nor sperm donors nor surrogate mothers could help this couple. Cloning, on the other hand, could give them a child that shared the genes of one of the parents.

Same-sex couples are another case where cloning makes biological sense. Without cloning, a lesbian couple could use a sperm donor to conceive, and a gay male couple could find a willing surrogate mom and egg donor. Both of these solutions, though, result in a child with only half of the genes of one of the parents in the couple. And both solutions involve a third party who is now biologically related to the child and may produce unwanted entanglements. A clone, on the other hand, is related only to one member of the couple. Especially for lesbian couples (who don't need the services of a surrogate mother), this may be the best option for having a biologically related child.

Reproductive cloning isn't safe today. Cloned animals are often born with deformities or health problems. So long as that's the case, it's unethical to attempt it with a human. The technique can and should be illegal until basic research on animals makes us confident that it will work safely in humans. However, the anticloning movement isn't primarily concerned with safety. It's concerned with humanity and identity, and the notion that giving birth to cloned humans would somehow rob them—or the rest of us—of our unique human dignity.

The debate about reproductive cloning is often marred by a misunderstanding of what it means to clone someone. In popular parlance, clone is shorthand for "copy." Imagine a clone of yourself. What do you see? You may be picturing someone about your age, who looks just like you, with your beliefs and attitudes, perhaps even your skills and knowledge of the world. But this would not be your clone.

In fact, there are 40 million human clones alive today. We call them identical twins. We recognize them as individuals without a second thought. We have no angst about whether a twin is less of a person than a singleton. We have no doubt that each twin has a soul, at least in so far as the rest of us have souls.

Clones conceived with the help of biotechnology would be just as much individuals as are naturally conceived identical twins. In fact, a clone of you would differ from you much more than a twin. Your clone would grow up decades later than you did, surrounded by different people and the culture of a different era. He or she would have a personality and beliefs different from yours.

There is no need to ban reproductive cloning on moral grounds, because clones would be ordinary people like the rest of us.

Genes and Human Behavior

One of the major concerns among those who oppose using genetic technology to select desirable traits of the unborn is the amount of power this could give parents over their children. Environmentalist Bill McKibben is among those who are concerned about this. In his book *Enough: Staying Human in an Engineered Age*, he writes: "The person left without any choice *at all* is the one you've engineered. You've decided, for once and for all, certain things about him. . . . You'll be taking this on your own shoulders, exercising infinitely more power over your child than your parents did over you. . . . He may have no more choice about how to live his life than a Hindu born untouchable."

How realistic is McKibben's concern? The answer to that question depends on many factors, and particularly on just how much our genes affect our identity.

It's possible to use genetic engineering to alter the odds that a child will have any given trait. Exactly how strong an effect you can have on the odds depends on what you're trying to change. Some traits, such as eye color, are almost completely genetic. Other traits, like height, are largely controlled genetically but also depend on the environment—in this case, nutrition. Most traits, however, depend heavily on both genes and environment. A child with the genetic predispositions of Arnold Schwarzenegger won't necessarily grow up to have Arnold's physique. Those muscles require both the right genes and long hours spent in the gym.

Behavioral traits are even harder to pin down, for they depend on a complex interaction of genes and environment. Some genes affect personality only in the presence of some environmental triggers. For example, Avshalom Caspi and collaborators at Kings College in London concluded in a 2002 study that children with a mutation in the gene for the brain enzyme monoamine oxidase A (MAOA) are more likely than average to commit acts of violence, but *only* if they're abused while young. Children with the MAOA mutation who grow up in a supportive, nonabusive home are no more likely to be violent than children without the mutation. Abused children who *don't* have the MAOA mutation are also un-likely to grow up to become violent. Neither the environment nor the gene produces an effect in isolation. They *both* have to be pres-ent to affect behavior.

Despite this complexity, we can make some general observa-tions of the degree to which genes affect various personality traits. These observations come from studies of families—particu-larly adopted children, twins raised in the same household, and twins separated at birth.

These studies work by comparing the scores of pairs of siblings and sometimes parents on a variety of tests. What they show is that genes have a substantial impact on personality, but not a dom-inant one.

For example, imagine two pairs of twins. One pair are identical twins—they have the exact same genome. The second pair are fra-ternal twins of the same gender. The fraternal twins only share half their genes with one another. So in a sense, the identical twins are twice as genetically related as the fraternal.

When researchers administer tests to these pairs, they find that the identical twins tend to have very similar scores, greater than the similarity between the scores of the fraternal twins. Since both pairs grew up together in similar environments, the greater genetic similarity of the identical twins is likely the cause of the greater similarity in scores.

In fact, numerous studies have shown that identical twins separated at birth have similar scores on a variety of personality and intelligence tests. On most of these tests, adopted children tend to resemble their biological parents more than their adopted parents. Overall the data suggest that genes play a large role in intelligence and personality.

But genes don't control everything. Identical twins are still quite different from one another and when raised apart from each other are just as varied as unrelated people of the same age in rates of smoking or church attendance or numerous other factors. Even on tests of intelligence and personality, identical twins separated at birth have significantly varying scores—they're just more similar to each other than the scores of unrelated people or fraternal twins.

So how large is the effect of genes on mental traits? Let's consider the effects of genes on tests of intelligence and personality.

The word *intelligence* means many things to many people, so researchers have instead focused on IQ. IQ tests look at simple problem-solving abilities, verbal skills, and spatial skills. They don't measure leadership abilities, depth of knowledge, motivation, creativity, emotional intelligence, interpersonal skills, or many other things you may think of when you call someone intelligent.

Despite that, IQ—the intelligence quotient—is a relevant number. Although IQ tests differ in terms of whether they're timed, whether they focus on verbal or spatial skills, how long they are, and other parameters, an individual's scores on all of these tests tend to be fairly consistent. What's more, according to a wide variety of studies, IQ is a fairly good predictor of educational achievement, of income later in life, and even in some cases of life expectancy. So if IQ tests don't measure the full extent of intelligence, at least they're measuring some things that we care about.

Researchers express the power of a gene over a human trait in terms of correlation, which is measured on a scale of 0 to 1. A correlation of 0 means the gene and the trait are completely unrelated. A correlation of 1 means that one trait perfectly predicts another. The correlation between genes and IQ is somewhere around 0.5. Most estimates have ranged between 0.35 and 0.75. This indicates that genes and environment play a fairly equal role in determining IQ. What this means is that the people with the highest IQs tend to be those with both the genes that encourage high IQ and an environment that encourages high IQ.

Personality is even more difficult to quantify than intelligence. The most agreed-upon model of psychologists is the division of personality traits into the so-called big five, five axes of personality on which an individual can be fairly reliably tested: neuroticism, introversion-extroversion, agreeableness, conscientiousness, and openness to experience. Like attempts to quantify intelligence, attempts to quantify personality in terms of these traits are controversial. There are several competing models, but the "big five" model is the most popular among psychologists of personality.

On the basis of twin studies, researchers estimate that the genetic contribution to these traits is somewhere between 0.3 and 0.5. Research from adoption studies produces a lower range of estimates, between about 0.2 and 0.35. So in general it appears that genetic variations account for about half of variations in IQ scores and for a third to a half of variations in personality test scores.

This level of correlation between genes and behavioral traits constrains the amount of power that genetic engineering will give parents to determine the characteristics of their children. It indicates that genetic engineering will give parents a tool that can increase their odds of having a certain kind of child, but that no amount of genetic tinkering will allow parents to completely control the IQ or personality of their child. The final personality, intelligence, and other behavioral traits of your child will still depend critically on environment and random chance.

To illustrate this, let's do a quick thought experiment. Suppose that you want to have a high-IQ child. You could increase the odds of this by giving your child the same genes as someone who's already lived and who had a high IQ. For argument's sake, let's say you pick Albert Einstein. Since this is a purely hypothetical case, let's also ignore all the issues involved in discovering his genetic makeup, determining which genes are involved in intelligence, and actually performing the genetic engineering.

We don't know what Einstein's IQ was, but it's been guessed at about 160. If Einstein did have an IQ of 160 and you conceive a child with the same intelligence-linked genes as Einstein, what's your child's IQ most likely to be? You may be tempted to guess 160, but remember that only about half of the variation of IQ from the norm is accounted for by genetic variations. The other half is a mixture of environment and chance. Einstein's hypothetical IQ differed from the norm by 60 points. So we should expect that about half of this variance, or 30 points, is accounted for by his genetic differences from the human norm.

That means that if we created a population of Einstein clones, and raised them in "average" environments, their average IQ would be 30 points higher than the norm, or 130. The individuals of the group would all have different IQs, of course, accounted for by their different environments and the random events in their lives. Those IQs would form a normal bell-curve–shaped distribution with 130 in the center. Fewer than one out of every four hundred of these clones would have an IQ of 160 or higher. Just as many would have an IQ of 100 or lower (see figure 9, page 158). Your odds of producing someone of average intelligence by copying Einstein's genes are just as high as your odds of producing an Einstein-level genius.

On the other hand, 99.8 percent of these children are going to be above the human average in IQ. More than 80 percent will have IQs of 120 or higher. And these numbers assume "average" environments. Copying Einstein's genes isn't likely to produce a super-genius, but it's very likely to result in a child with a substantially above-average IQ.

This example illustrates both the power and limitations of genetic engineering. Children carrying genes linked to high IQ are going to have higher average IQs than the norm. But the only ones to reach truly high IQs are going to have environments and random life events just as exceptional as their genes.

To be fair, we've played a bit fast and loose with genetics in this thought experiment. The truth is that we don't know Einstein's IQ for certain. Even if we did, we wouldn't be able to say how much his high IQ stemmed from exceptional genes and how much from an exceptional environment. The correlations between genes and IQ apply to a *population* and have little to say about a given individual. Perhaps Einstein's environment was more exceptional than his genetics—or vice versa. Still, the overall point holds, and we can state the thought experiment a bit more precisely. If you took a large number of individuals with 160 IQs, cloned them, and raised the children in randomly selected homes, the average IQs of the children would be 130.

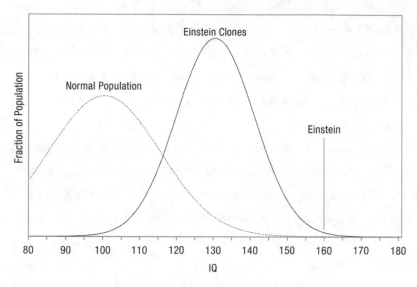

IQ Distribution of Einstein Clones

FIGURE 9 Only 1 in 400 of Einstein's clones raised in an average environment would have an IQ equal to or greater than Einstein's IQ.

Exceptional individuals usually have both exceptional genes and exceptional environments.

Personality measures follow the same basic pattern. For example, selecting genes associated with a high degree of openness to experience increases the likelihood that you'll get a very open child, but doesn't guarantee it. And genetic alterations of personality come with another risk: of overshooting the target. If you genetically engineer an embryo to select genes associated with, say, agreeableness, you increase the odds of getting not only a pleasant, agreeable child but one who is agreeable to a fault. If you genetically engineer your child to increase the odds that he'll be an aggressive go-getter, you also increase the odds that he'll become an overbearing boor.

As linguist and evolutionary psychologist Stephen Pinker puts it:

Most traits are desirable at intermediate values. Wallis Simpson said that you can't be too rich or too thin, but other traits don't work that way. Take aggressiveness. Parents don't want their children to be punching bags or doormats, but they also don't want Attila the Hun either. Most want their children to face life with confidence rather than sitting at home cowering in fear, but they don't want a reckless daredevil out of *Jackass*.

The final limitation on genetic control of personality is the fact that any given gene plays multiple roles. A gene that affects one facet of personality or intelligence is likely to affect others. For example, variations in the COMT gene are believed to account for 4 percent of human variation in IQ. COMT comes in two varieties, "val" and "met." Everyone is born with two copies of the gene. People born with two copies of the "met" version of the gene seem to make more efficient use of the frontal lobes of their brain. They tend to have better working memories, and perform better than other people on various tests. Yet they also seem to be more prone to anxiety, more sensitive to pain, and more likely to develop schizophrenia than people with two copies of the "val" version, or one each of "val" and "met."

As we start to decipher the role of genes in our behavior, we'll find that many and perhaps all play multiple roles, so choosing specific genes is never an à la carte order; there will always be trade-offs. Perhaps a parent will have to decide if an increased likelihood of high IQ is worth the increased chance of anxiety or mental disorder. Thousands of other genes will come with such trade-offs.

These trade-offs and other limitations shed light on another concern about genetic engineering: the worry that power over our children's genes will homogenize society. Few parents with access to genetic technology are going to choose to have children with bubble boy disease or muscular dystrophy or a predisposition for heart disease or cancer. But among the behavioral traits, the trade-offs are so numerous and control so imprecise that parents are bound to bring a wide menagerie of children into this world.

Genes exert a tremendous impact on our behavior, and in the near future parents will be able to employ genetic engineering to influence their children's health and appearance: hair color, eye color, skin tone, height, and facial structure; health factors, athleticism, and weight. They'll also be able to influence, but never completely control, their child's IQ, personality, and other mental traits. Genetic engineering will give parents substantial choice regarding their children's personalities, but far from absolute power.

This is not to say that cloning and genetic engineering won't cause any problems. In 2001, Margaret Talbot wrote an article on cloning that appeared in *The New York Times Magazine* that illustrates one realistic concern. Talbot talked to parents who had lost a child and were interested in cloning their deceased son or daughter. The parents all claimed to understand that two children who had identical genomes would nevertheless be unique individuals—that nothing could bring their son or daughter back to life. Despite this, they seemed to cling to an emotional view of cloning as a way to resurrect their lost child.

A clone will never be the same as the dead child he or she is cloned from. Parents who think they can bring back their lost son or daughter are bound to be disappointed, and a child always measured against an invisible yardstick of someone else's life is going to suffer.

Similar issues may arise with genetic engineering. A failed athlete may engineer his son or daughter for physical abilities and push that child into sports whether the child likes it or not. If the child fails to excel, the father's reaction may range from severe disappointment to emotional or even physical abuse. An academic may engineer her son or daughter for extreme intelligence but react in dismay if the child chooses a career in art rather than science.

Even with good parenting, the emotional consequences for a child who has been "engineered" might be negative and severe. Children who know that they were "designed" to have a certain trait may feel deeply resentful, as if their lives have been controlled and scripted from the outset, leaving them no chance for individuality. They may even resent their parents for not engineering them as they would have wished: providing them with great athletic abilities when what they really want is to be an intellectual, or being unwilling to sacrifice financially to buy them better looks or a longer life span.

This is not a new concern. When the first babies conceived via in vitro fertilization were born, many critics of the technique claimed that these children would be emotionally scarred for life, either by the process of gestating for a few days in a test tube or, more likely, by the very knowledge that they are "artificial" children, born in an "unnatural" way. Biotech gadfly Jeremy Rifkin warned about the "psychological implications of growing up sheltered not by a warm womb but by steel and glass, belonging to no one but the lab technician who joined together sperm and egg." Even as recently as 2002, Leon Kass warned that any conception that occurs in a laboratory is "profoundly dehumanizing, no matter how genetically good or healthy the resultant children."

Other bioethicists warned of other psychological problems

for IVF children. Some pointed to the stress of past failures to conceive or of the IVF process itself as a reason why IVF families might be weaker. Others suggested that an infertile couple who placed all their dreams in IVF might conjure up a "fantasy child" that no real child could live up to.

Yet psychological studies years later show that IVF children have grown up to be perfectly normal and well adjusted. In fact, if anything, they tend to be *more* emotionally healthy than the norm. The data suggest that parents who went to the effort to use in vitro fertilization tended to provide a healthier upbringing for their kids, on average, than parents who conceived naturally.

In a 1998 review published in the journal *Biomedical Ethics,* researchers Emma Goodman, Fiona MacCallum, and Susan Golombok summarized data from all the studies on IVF families published to date and concluded that "the children themselves seem to be developing normally" and that in IVF families "the quality of parenting appears to be superior, with IVF mothers displaying higher levels of warmth and emotional involvement, and both parents interacting more with the children than in naturally conceived families."

That may reflect the fact that parents of IVF children obviously wanted to have those kids. They committed hours or days researching the technique and visiting doctors, and ultimately underwent surgery and spent thousands of dollars to conceive their child. It's reasonable to expect that they'd be committed and involved parents as well.

The same logic extends to parents who—going a step beyond IVF—genetically engineer their children, especially while the techniques are still fairly new and expensive. These parents are molding their child, not just trying to bring him or her safely into the world, but trying to pass on what advantages they can. Parents who go to this extra trouble and expense *care*. These are the kinds of parents that are likely to be active in their children's lives and produce healthier and better-off kids for it, even without the genetic advantages they have conferred.

Investment in the Future or Eugenics?

Genetically engineering our children is a new idea, but investing in them is not: over the course of human history, parents have always sought to give their children whatever advantages they could, to feed, clothe, house, and teach their children as well as they were able. Those advantages have in many cases altered the brains and bodies of those children for life.

Indeed, parental investment in children is selected for by evolution. Any parent who doesn't pour energy into giving his or her children the best possible place in life is going to be at a disadvantage in passing genes on to the next generation. Unlike animals that can have dozens or even thousands of offspring in a generation, we humans have only a few, and they take longer than the children of any other species to grow to maturity. As a result, we're even more evolved to care for, nurture, and provide advantages to our children than parents of other species.

Our societies welcome this, celebrate it, and build institutions around it. Our schools, vaccination programs, piano lessons, private tutors, and more are all attempts to pass on whatever benefits we can to our children. As bioethicist Arthur Caplan notes: "The notion of enhancement is deeply ingrained in our culture. The only debate about the Kaplan SAT prep test is how many weeks to attend. What about spending $25,000 a year to send your kid to private school? Nobody says, 'You should be ashamed, you're giving your child an advantage.'"

Schooling, nutrition, and other cultural phenomena can make a lasting difference. Improved nutrition is credited for increasing average height of men and women in the United States by around three inches since the colonial period. Similar changes have occurred in other countries around the world.

In 1984, political scientist James Flynn noticed another interesting trend, this time in human IQ. IQ scores are normalized to 100, meaning that no matter how well people perform, the average performance in a given year is assigned a score of 100, and the other scores are distributed on a curve around it. For that reason, it's easy to miss something deeper going on. Flynn took the raw score on IQ tests, the number of correct and incorrect answers, and found that the number of correct answers had been increasing steadily over time. In the United States, Flynn saw a 25-point gain in IQ test results between 1918 and 1995. In other words, if the tests of people who scored 100 in 1918 were scored against the 1995 curve, they'd result in an IQ score of 75. The 1995 tests that scored 100 would score 125 on the 1918 curve.

Flynn went on to find the same effect in multiple intelligence tests from different countries over different periods of the twentieth century. The test results from the United States were on an IQ test called the Stanford-Binet. In West Germany, Flynn found, IQs rose by 20 points between 1954 and 1981 on a different test called the WISC. In France, results rose 20 to 25 points between 1949 and 1974, according to Flynn's analysis of a test called Raven's Progressive Matrices. Worldwide, the average rise has been around 3 points per decade, with results to back this from France, Germany, Japan, the Netherlands, the U.K., Greece, and dozens of other countries. Interestingly enough, the best-documented gains are not on tests that measure concrete knowledge of vocabulary and facts, or crystallized IQ. Instead they're on tests of "fluid" IQ: abstract reasoning, spatial skills, and visual skills.

The "Flynn effect" can't be put down to genetics because of the rapidity with which the phenomenon has occurred. Flynn was able to demonstrate this by documenting a large increase even between fathers and their own sons. There is active debate as to exactly what causes the effect, but most speculation centers around better nutrition, better schooling, and increased exposure to information of all sorts during childhood.

Both increasing height and improved IQ show that children of recent generations have been enhanced through nongenetic techniques. Yet at the end, improvements in nurture are as effective and solid as improvements in nature. Whether a child grows up to be taller or more mentally nimble as a result of genetic changes or environmental ones doesn't really matter to the child or the parent. What does matter, to almost every parent, is to do the best we can for our children, affording them the advantages we can, as nature has engineered us to do.

For both cultural and biological reasons, it's a basic tenet of society today that parents are the best people to make choices on behalf of their children until the children are old enough to choose for themselves—unless the parents demonstrate obviously destructive behavior. We trust parents to choose the schools children go to, the church they attend, the language they learn, the food they eat, the television and movies they watch, and more. This trust reflects the fact that for the most part, parents want what's best for their children and are in a better position to look after the best interests of their offspring than anyone else.

It's interesting, then, that some opponents of genetic enhancement have labeled the practice "eugenic." For example, Leon Kass, the chairman of the President's Council on Bioethics—who has made it clear that he would like to see a ban on genetic engineering—has written that the debate over cloning offers a unique practical opportunity to combat "the impending practice of germline genetic manipulation or other eugenic practices." Boston University bioethicist George Annas, who's proposed a UN ban on genetic engineering, went so far as to say "modern genetics is eugenics," while visiting the Holocaust Museum in Washington, D.C.

Gilbert Meilander, another member of the President's Council, wrote in a 2001 essay: "Our present condition is this: we have entered a new era of eugenics. That science which attempts to improve the inherited characteristics of the species and which had gone so suddenly out of fashion after World War II and the Nazi doctors now climbs steadily back toward respectability."

The connection to World War II Germany here is quite explicit. The use of the word *eugenics* is intended to form an association in the readers' minds between any attempt to influence the genes of one's child and concentration camps, genocide, gas chambers, and Nazi medical experiments on humans. While "eugenics" may at one point have referred to efforts to improve humankind, today the association is with attempts by totalitarian states to control the makeup of the population, usually at the barrel of a gun.

What's interesting is that in the debate over human enhancement, the only people advocating state control over the genetic makeup of the population are those who would like to see genetic enhancement techniques prohibited. The advocates of human enhancement, on the other hand, are arguing for individual and family choice, the opposite of state control.

To a certain extent, then, the prohibitionists are the ones upholding the eugenic side of this debate. It's those who oppose individual and family genetic choice who have, in essence, decided that there's a certain "correct" genetic heritage for humanity (the one we have today) and that the populace should not be allowed any choice in the matter. The relatively small number of advocates for genetic choice, on the other hand, are not trying to impose their opinion on the rest of the country or the rest of the world. They're only seeking the freedom for people to make genetic decisions for themselves and their families—the same freedom they enjoy to make the thousand other decisions involved in raising a child.

Regulating Reproductive Technologies

A regulatory regime consistent with family choice would focus on safety, education, and equality rather than prohibition. Today, regulation of reproductive technologies varies by country and by the specifics of the technology. The most consistent international regulation applies to reproductive cloning. More than thirty countries, including Russia, Japan, China, and most of Europe, have banned reproductive cloning. In 2003, two efforts to ban cloning, in the United States and internationally through a UN agreement, both failed, despite widespread support for a ban, owing to disagreement over whether bans should also cover therapeutic cloning for the production of stem cells. Nevertheless, reproductive cloning will likely be illegal throughout virtually the entire world within a few years.

At the other end of the regulatory spectrum is in vitro fertilization, which is legal in most countries. In some countries (Japan, Australia, parts of Europe), the cost of the procedure is subsidized by the government.

Neither genetic engineering of human embryos nor in utero

gene therapy has ever been attempted (as far as we know), and so the legality of these procedures in most countries has not been tested. A few countries, such as Germany, already have laws on the books that ban such procedures, but most do not.

The most interesting case—and perhaps the best view into the future—is preimplantation genetic diagnosis (PGD). PGD laws differ quite a bit from country to country. In the United States, PGD is legal and is virtually unregulated. Professional standards for physicians mandate follow-up studies, but there's no regulatory agency whose approval is required before performing a procedure. Nor is there any agency whose approval is required prior to the introduction of entirely new medical techniques aimed at improving reproduction.

PGD is also legal in China and India, though both countries outlaw its use as a means to select the gender of a child. In Japan, the technique falls outside the guidelines of the Obstetrics and Gynecology Society, but it isn't prohibited by law. At least one doctor is already providing PGD as a service. In Australia, the technique is legal in most states but there are slight variations in the criteria for when it may or may not be used. In Europe, laws on PGD range from fairly permissive (Belgium) to complete prohibition (Germany). In England, where PGD was first performed, every couple seeking to use the technique (or any new reproductive technique) must first apply to the Human Fertilisation and Embryology Authority. The HFEA approves requests whose goal is to produce a child for an infertile couple or to avoid passing a clearly demonstrated hereditary disease down to a descendant, and rejects almost all others.

The inconsistency of laws between countries reflects local cultural values but has international consequences. It's clear that already parents are crossing national borders to obtain PGD. For example, a 2003 study indicated that roughly two thirds of the couples receiving PGD services in Belgium, which has Europe's most permissive PGD laws, come from other countries. There have also been highly publicized cases of British couples who, after having been turned down in their request for PGD at home, traveled to the United States to get the service.

Clearly, unless laws covering genetic enhancements are consistent between countries, they won't stop motivated parents. Given the attitudes toward genetic technologies in different countries, it's unlikely that the laws will ever be fully consistent. As we observed earlier, adding the burden of travel also increases the cost of a service, tilting access away from the poor and toward the rich.

This is not to say that the field should be unregulated—it should be regulated in a way that makes technologies available to the public and emphasizes safety, education, and equality. The United States in particular needs more regulation than it has. At a national level, there's no law or regulatory agency that serves as a gatekeeper for new reproductive techniques or evaluates their safety before they're attempted. It's fortunate that both IVF and PGD (which were developed under similar conditions in England) have turned out to be quite safe and have produced healthy children, but there's no legal or scientific review process in place to ensure the safety of the next innovation in reproductive medicine. Prior to their first use in humans, reproductive techniques ought to go through the same sort of preclinical trials the U.S. Food and Drug Administration requires for drugs and gene therapies.

At the point at which reproductive techniques *are* shown to be safe, the role of government should be not to deny families access but to promote education and equality. Education in this area is particularly challenging because of common misperceptions about clones and genetic engineering. Any educational campaign has to focus on communicating the fact that every child is an individual and that genetic techniques cannot guarantee a particular personality, intelligence, degree of health, or even appearance.

As a practical matter, a regulatory model grounded in individual and family choice to employ genetic technologies isn't likely to be adopted anytime soon. Reproductive cloning, in particular, is likely to become illegal throughout the world and remain that way for the foreseeable future. The flashpoints of debate for the next decade will be the techniques that are regulated inconsistently or not at all today: PGD, in utero gene therapy, and genetic engineering.

To a certain extent, so long as these techniques remain legal in at least one major country, the debate will remain alive. In particular, support for the medical uses of genetic technologies is high in the United States and Asia, and the precedent of Europeans going abroad to seek out reproductive services has been set. And over time, forays into the gray area between therapy and enhancement may acclimate the public to the use of genetic technologies to alter a wider set of traits.

ULTIMATELY, WHATEVER CHOICES we may make for our children will be subject to change, at their choice, when they reach adulthood. In the coming years, pharmaceuticals, adult gene therapy, and the integration of computers into the brain will give people far more control over their own minds and bodies than we enjoy today.

Genetic engineering will give us the power to influence the initial nature of our children, but that nature will become more and more malleable over time, as new technologies that allow our children to modify their adult minds and bodies mature. In adulthood, our children will be able to choose who they are, regardless of how we conceive and raise them.

CHAPTER 9

The
Wired
Brain

I N N O V E M B E R 1 9 9 7 , Johnny Ray, a fifty-three-year-old Vietnam veteran, seemed perfectly healthy. He spent his days as a drywall contractor and his evenings and weekends playing blues guitar. He had no reason to expect a major health problem, yet everything in his life changed that month when he suffered a massive stroke.

Ray was on the phone at his home in Douglasville, Georgia, when it happened; weeks later, he woke up in a different world. He was in a hospital bed in Atlanta, at the local Veterans Affairs Medical Center. He was completely paralyzed from the neck down. A tracheotomy had been performed to save his life—and it had taken away his voice. He could hear, think, and understand, but he could not act. He could express himself only by blinking his eyelids: once for "no," twice for "yes."

Every day in the United States, two thousand people suffer strokes. More than seven hundred thousand have one each year. Other problems like ALS (Lou Gehrig's disease) and neck injuries take their own toll. There are at least a quarter million quadriple-gics in the United States. Worldwide half a million "locked-in" pa-tients like Johnny Ray suffer more than just neck-down paralysis: they're unable to eat, drink, or talk on their own. They can ob-serve the world but not interact with it. For them, virtually any ability to express themselves would be an immense improvement in their quality of life.

Just across Atlanta from the Veterans Affairs Medical Center is Emory University, where a neurologist named Phil Kennedy had been working on an implantable device to restore some degree of freedom to locked-in patients like Johnny Ray. Over the years, his research with monkeys had shown that an electrode implanted in the brain could facilitate communication to the outside world, by picking up a brain signal and transmitting it to a computer. But no one had ever attempted such a procedure on a human being.

Kennedy is a tall, patient Irish gent who's been transplanted into the American South. He's modest and careful with his words, while radiating a kind of solid determination. Kennedy believed his technology could help patients like Johnny Ray, and on the ba-sis of his animal studies, he was able to make a persuasive case to the FDA for permission to conduct a human trial. Crucial in his proposal was that the electrodes would be wireless. After the sur-gery, no wire would protrude through the skull into the brain, limiting the risk of infection. And as for the other risks of the sur-gery, the patients were willing, for they had very little left to lose, and a tremendous amount to gain. The FDA approved.

In March 1998, neurosurgeon Roy Bakay implanted the device into Ray's motor cortex—the part of the brain that deals with mo-tion. Specifically, he implanted it into the area of the motor cortex responsible for moving the left hand. The cortical control system, as Kennedy calls his implant, had just a single electrode. It seemed almost impossible that it could pull out a useful signal from the hundreds of millions or even billions of neurons in a human brain involved in motor control.

Nevertheless, animal studies had suggested it was possible. Johnny Ray was put through an intensive training program. He would stare at a computer monitor that showed an on-screen keyboard and had a thought-controlled cursor with which to pick out letters. Under Kennedy's guidance, Ray would think about moving his left hand: up if he wanted the cursor to move up, down if he wanted the cursor to move down, and so on. As he imagined moving his hand, the electrode in his brain picked up the signals of the few neurons near it and broadcast those to a nearby computer. The computer in turn moved the cursor. Something amazing was happening. A human was just *thinking* about something and a computer was responding. It was the stuff of science fiction.

Learning to move the cursor wasn't easy, though. The electronic pointer didn't always go where Ray wanted it to go. Gaining control over it was like learning to move his hand all over again, an exhausting trial-and-error process. Yet little by little, Ray got the hang of it. Several months later, he was typing out names and sentences, and clicking icons to say things like "I'm hungry." What's more, by Ray's reports, he had *stopped thinking about moving his hand*. He simply focused on the letters and icons he wanted to click. He willed the cursor to move without any kind of intermediary. In some sense, the computer had become a part of him.

Today there are millions of people worldwide who lack some basic human ability that could be restored by a brain-computer interface—blind people who could see again, deaf people who could hear again, paralytics like Johnny Ray who could communicate and move again, and those with severe brain damage who could think again.

Researchers have already made dramatic progress in assisting patients with these problems. Over the last few years, experiments have restored sight to the blind and hearing to the completely deaf. Brain implants have been used to control the tremors of Parkinson's disease and the seizures of epilepsy. Animal studies have produced the sensations of touch, pleasure, and pain. Current work is looking at restoring damaged memories.

As we now know, the quest to heal often leads to the power to enhance. As we learn how to repair damaged brains, we'll discover an immense amount about how the brain works. That in turn will lead to devices that can improve our mental abilities.

Using electrodes and brain chips, scientists already have learned a lot about how the brain encodes three of our senses: sight, hearing, and touch. They have learned which neurons in the human brain fire in response to which kinds of images, sounds, and sensations. In some sense, we've begun to crack the code of these functions of the brain.

The next frontier is to discover how the brain encodes our higher mental functions: language, memory, attention, and emotion. We now have the technology to study the brain and discover exactly these things. If we can learn these neural codings, if we can crack the code of our highest mental functions, we can manipulate them. We can repair them when damaged. We can, potentially, improve on them—giving ourselves better memories or new mental abilities. We may be able to control them—giving ourselves the ability to alter our feelings, thoughts, and perceptions. We may even be able to transmit some of the private contents of our brains—our thoughts, memories, and emotions—directly from one person's brain to another's.

So far that's science fiction. We have just begun to take our first forays into repairing damaged brains and transmitting information in and out of them. It will take years of painstaking research to decipher the neural codes of the higher brain functions. Along the way there will be mistakes, human trials that harm rather than help their patients. Even once neural prosthetics are routinely employed to repair damaged brains, it will require a revolution in surgery before implanting them is safe and convenient enough to entice perfectly healthy men and women to receive these implants.

Neural interfaces are worth discussing not because they're right around the corner—they're not—but because they may effect such tremendous changes in our lives. The ability to tinker directly with the inner workings of our brains and connect them directly with computers offers us more power over ourselves than any technology we've discussed thus far. It offers us the potential to control our emotions in real time, to rescript our personalities, to communicate our innermost thoughts and feelings to one another, even to draw on the native abilities of our computers as if they were our own. These abilities will pose serious questions to our sense of identity and individuality. They will blur the line between man and machine. They may even blur the line between one person and another. More than genetic engineering of our children, more than the ability to genetically and pharmacologically alter our minds and bodies, more even than the conquest of aging, the integration of our minds with our computers and through them with each other will challenge our ideas of what it means to be human.

For now, medical research into using brain implants to assist the blind, the paralyzed, and those who've suffered brain damage will drive the development of the technology. The urge to help and heal will propel the design of more sophisticated implants and safer surgical procedures to implant them. It will teach us more about how the brain works, and it will gradually introduce the idea of brain implants to the general public. This will blunt the shock that will come when people begin using these implants to enhance themselves rather than cure disease. With this path in mind, it's time to look more closely at the history of brain—computer-interface research and what it tells us about the future.

Robotic Limbs

Research on restoring motion to the paralyzed didn't stop with Johnny Ray; Phil Kennedy and his collaborators have implanted electrodes into the motor cortices of another dozen patients since then. Kennedy's latest studies are using eight electrodes rather than just the one that Ray received, allowing their patients to learn to use the on-screen typing system more quickly and easily. Animal studies have gone even further, and research in the area has been boosted by funding from an unexpected source—the military.

DARPA, the Defense Advanced Research Project Agency, is the U.S. military's premier science agency, charged with funding and conducting basic research that will give the U.S. military a cutting edge in battle and that will also find its way into civilian use. Odds are you've already benefited from a DARPA research project—the Internet is an outgrowth of something called ARPANET, which DARPA's predecessor agency created as an experiment in computer networks that could survive a nuclear war.

In 2002, DARPA gave $24 million to a team of researchers led by Miguel Nicolelis of Duke University and John Chapin of the State University of New York. Nicolelis and Chapin are working on brain-computer interfaces to give the paralyzed control of robot arms. DARPA's interest is in giving U.S. soldiers the ability to control tanks, fly planes, and share information purely through thought. By cutting the hand out of the loop, DARPA hopes that reaction time can be improved and that pilots will gain the ability to control more complex systems than their hands can currently manage.

Miguel Nicolelis came from Brazil to Hahnemann University in Pittsburgh in 1989 at the age of twenty-eight, with both an M.D. and a Ph.D. from the University of São Paulo under his belt. He's a serious-looking man with a thick bushy beard, receding hair, and sad eyes behind large square glasses. He'd always been interested in decoding the function of the brain. At Hahnemann he met John Chapin, already ten years into his own career and working on ways to record the activity of neurons.

Over the fifteen years of their collaboration, Nicolelis and Chapin have produced some of the most impressive results in the field of direct brain control of computers and robots. They've managed to increase the speed and accuracy of using a computer prosthetic beyond what Kennedy achieved, while dramatically reducing the amount of training time the patient requires.

Their first breakthrough was in 1999, when they authored a pivotal study that showed that a rat could control a robot arm in much the same way that Johnny Ray could control a computer cursor. In 2000, they showed that an owl monkey could do the same, and added a new twist. Instead of sending the signals from the monkey's brain straight to the robot arm, they transmitted the information about the monkey's neural activity across the Internet to control a robot arm in a lab at MIT, six hundred miles away, suggesting that humans with neural prostheses could control robots in remote hazardous environments or pilot unmanned planes over a battlefield from the comfort and safety of a base.

In their 2003 study, funded by DARPA, the researchers implanted seven hundred electrodes in the motor cortices of rhesus monkeys, which are close relatives of humans, and trained each monkey to control a robot arm using a joystick. As the monkey used the joystick to control the arm, a computer recorded the activity of its neurons. Using a special program built for the task, a computer attached to the electrodes found a pattern in how the neural firing related to the movement of the robot arm. The researchers had cracked the neural encoding of motor control. Armed with this code, they could now predict how the robot arm would move in response to the activity in the monkey's motor cortex. In trial after trial, their predictions matched what really happened.

Then came the moment of truth. Nicolelis and Chapin left the joystick in place but disconnected it from the robot arm. They gave the computer that monitored the monkey's implant control over the robot arm instead. Now the motion of the arm would depend entirely on the signals picked up by those seven hundred electrodes, and not the way the monkey moved the joystick. Would the code work? It did. The monkey was now moving the robot arm purely by thought.

In fact, their processing of the code was a bit too rapid, and the robot arm moved before the monkey expected it. While nerve signals move at a relatively slow 100 meters per second, signals in electronic circuits travel at close to speed of light, about three thousand times faster. So not only had Chapin and Nicolelis predicted the way the monkey would want to move the robot arm on the basis of its neural activity; they had cut out the delay of sending those signals from the monkey's brain to the muscles that controlled its hand.

Something even more surprising happened next: in some trials, the monkey stopped using the joystick at all. It let its own hand drop to its side and relied solely on the neural interface to control the robot arm. According to Nicolelis, "The monkey suddenly realized that she didn't need to move her arm at all. Her arm muscles went completely quiet, she kept the arm at her side and she controlled the robot arm using only her brain and visual feedback. Such findings tell us that the brain is so amazingly adaptable that it can incorporate an external device . . . as a natural extension of the body."

Such a natural extension has interesting consequences. Before the implant, the monkey had four limbs: two arms and two legs. Now it had control of five limbs: two legs, two biological arms, and one robot arm. This is the kind of thing that gets DARPA excited. It suggests that a pilot with an advanced brain implant might be able to do more things at once than a pilot using just two arms and two legs. Human attention, rather than our number of limbs, may very well be the bottleneck. But the very hint of an advantage has DARPA interested in follow-up studies.

Nicolelis and Chapin's work incorporates three other advances over the implant that Johnny Ray received. First, tapping into the activity of more neurons increases the amount of information coming out of the brain. It increases the system's bandwidth: the speed and strength of the connection between brain and computer. By using hundreds of electrodes instead of just one, Nicolelis and Chapin were able to give the monkey much quicker and more accurate use of the robot arm than Ray achieved with his cursor. Second, the new robot arm required very little training.

The monkey learned to use it in about two days, as opposed to the months of training that Ray went through. This is partially a result of the higher bandwidth between the monkey's brain and the computer. It's also a result of a computer program that did some of the learning itself, rather than relying entirely on the brain to adapt. Finally, in previous experiments, Nicolelis and Chapin showed that they could also send touch information back into an animal's brain. That's a necessary step for a useful prosthetic arm in humans.

In 2004, the researchers tested a version of their system on humans undergoing brain surgery to receive an implant of a device to control Parkinson's disease. Nicolelis and Chapin had a window of just five minutes with each Parkinson's patient to test their techniques; in that time, they implanted a temporary device to monitor the patient's neural activity, and then showed each patient how to play a simple video game using a computer joystick. While the patients played the game, the device's array of thirty-two electrodes analyzed their patterns of brain activity and software tried to correlate their neural signals with the motions their hands were making on the joystick. Amazingly, after just a few minutes each, the system was able to find a pattern, offering the best evidence yet that a high-bandwidth neural interface could immediately assist paralyzed persons.

Nicolelis's team has received FDA approval to run more experiments on humans. At the same time, their success has sparked interest and competition in the field. More than a dozen universities and at least three private companies are now working to develop the brain-computer interface; many of them have produced results almost as impressive as Nicolelis and Chapin, and several of the teams plan human trials in the next few years.

Restoring the Senses

Before Phil Kennedy implanted an electrode in Johnny Ray, William House was working on restoring lost hearing. In the 1960s, House placed the first implants in the cochleas of profoundly deaf patients, restoring them to partial hearing.

Normal hearing aids work by amplifying sound that they pick up and then relaying it acoustically to the inner ear. In other words, they turn up the volume on the outside world. There are 34 million deaf or hearing-impaired people in the United States, and for most of those people, this sort of hearing aid can help. However, for the 2 million Americans who've lost the hair cells in the inner ear that stimulate the auditory nerve, turning up the volume does not help. In these people, there is nothing left in the ear to pick up the sound vibrations and turn them into neural impulses.

Today, more than seventy thousand patients worldwide have cochlear implants. A modern cochlear implant is a microphone with twenty-two electrodes, which electrically stimulate the auditory nerve. The different electrodes respond to different frequencies of sound. The more sound an implant picks up in one of those ranges, the more rapidly the corresponding electrode fires. For example, when an implant microphone picks up extremely low-frequency noises, it triggers the first electrode. The higher the volume of the low-frequency noise, the more rapidly the low-frequency electrode fires.

The auditory nerve that cochlear implants stimulate has more than thirty thousand fibers, each of which can carry a separate signal. A single electrode can stimulate somewhere between one and ten such nerve fibers. Knowing this, no one would expect that a twenty-two-electrode implant would be able to produce useful hearing, but it does. Cochlear-implant patients can generally pick up about 90 percent of two-syllable words by hearing them alone, and more meaning from context and by watching the lips of the speaker. This isn't perfect hearing, or even very good hearing, but it's a marked step up from deafness. It's also surprisingly good considering the tiny number of electrodes compared to the high number of nerve fibers.

We'll see this again in every other kind of neural interface we look at: just one or a few electrodes is enough to produce useful results, even when the brain area to which the electrodes connect contains tens of thousands or even tens of millions of neurons. In part, we have to credit the amazing adaptability of the brain for this. The human brain seems remarkably good at interpreting and pulling information out of any regular signal. As we'll see, this quality of the brain is a major benefit to brain–computer-interface research.

Patients who have suffered hearing loss from damage to the auditory nerve itself cannot be helped by cochlear implants, but research is ongoing into implants that would send signals directly into the auditory cortex of the brain rather than through the auditory nerve. Initial results seem promising, though the technology is far from being ready to market to the public.

Cochlear implants demonstrated for the first time that electronic devices can be usefully integrated with the brain, even with only a single electrode. Building on this success, other researchers have gone on to work on restoring and improving the other senses, especially vision.

In September 2002, CNN showed a video clip of a man named Jens Naumann, a fit-looking thirty-nine-year-old Canadian who chops wood and plays the piano, driving a Mustang convertible around in an office parking lot. The only unusual thing about the video was Naumann himself—he's blind, and had been for nearly twenty years. That was before he met Dr. William Dobelle.

There are 1.3 million blind people in the United States, and millions more worldwide. Dobelle has been working for thirty years to build a system that can restore sight to at least some blind people. He does this by implanting electrodes in the visual cortex of the brain and connecting them to a computer. The computer decodes video signals from a camera mounted on the patients' glasses and turns them into electrical impulses to send down the electrodes and into the neurons of the brain. Dobelle's patients don't get perfect vision, they don't even get very good vision today, but what they get is enough to be useful—it was enough for Naumann to drive that Mustang around without hitting the lamp poles in the parking lot.

Dobelle's announcement of the successful implantation and the subsequent press attention caused an uproar in the field—not because of his research per se but because of the way he went about it. Unlike the vast majority of brain–computer-interface researchers, Dobelle is not in academia. He seldom publishes results, preferring to keep his methods secret in the hopes of patenting them. He circumvents the normal medical and ethical review boards and FDA approval by having the surgery to implant his devices done outside the United States. And unlike other trials of experimental techniques, which are typically free to patients and supported by government grants, Dobelle charges patients for the procedure, which he says costs around $100,000.

Still, it's hard to argue with Dobelle's results. He's given us a remarkable proof of principle. What we've learned is the profound and important result that *it's possible to stimulate the brain to produce useful vision.* If other researchers can improve on Dobelle's work and bring the costs down, such techniques might someday be used to cure blindness in millions of patients.

The focus of visual prosthetic research today is restoring normal vision, but everything about the technology lends itself to someday providing *super*normal vision. The implant Dobelle inserted in Naumann's brain gets its signal from an external video source. Right now, that video comes from a camera worn by the patient. In principle, though, the video can come from any source to which the patient has access. To the computer and the electrodes in Naumann's brain, it's just another video signal, regardless of where it comes from—a computer, a DVD player, a video-game console, a virtual reality system, a remote surveillance camera, or maybe even a camera worn by someone else. Those video devices could themselves have abilities beyond those of the human eye. Want digital zoom? Dobelle built this feature into his first visual prosthesis, for a patient named Jerry. Want infrared or X ray? Want a visual representation of sonar? All of these are possible for a human with a digital video input interface.

The most important thing Dobelle has done is start to crack the code of vision, in much the same way that Kennedy, Chapin, and Nicolelis have made progress in cracking the neural code of motion. Prior to Dobelle's work, scientists had a general idea of how the brain encodes sight. Now we have confirmation that we know the code well enough to put a crude image into someone's mind. That code is now a kind of video format. Today we already have different video formats for television, digital television, DVDs, and several different types of computer video. Now "neural video format" has been added to that list. Given an image you'd like someone to see, Dobelle can figure out what pattern of electrical impulses to send into the visual cortex. He can do the same with a whole series of images—say, the frames of a movie. Just as we can now convert from any electronic format to any other, so we'll be able to convert any video that exists today into this neural video format.

Dobelle isn't the only researcher working on visual neural interfaces; at least half a dozen groups around the world are pursuing this goal. The field is split between two general techniques: those that connect a computer directly to the brain and those that connect to the retina. Dobelle is in the direct-to-the-brain camp. His implant sits on the surface of the brain area called the primary visual cortex, which is the first of several brain layers that perform visual processing. It is located at the very back of the brain, conveniently near the skull and accessible to surgery. Another convenient feature of the primary visual cortex is that it encodes visual information in a pretty simple manner: patterns of neural activity in this area tend to be shaped like the object you're looking at. If you focus on a large red triangle, a triangular area of your primary visual cortex becomes more active. This kind of simple coding lends itself well to interfacing with an external device.

Retinal interfaces are similar in some ways to cochlear implants. Like the latter, retinal prosthetics are not connected to the brain; instead, they're wired to the retina, the layer of cells at the back of the eye that pick up light, or to the optic nerve, which carries information from the eye to the brain. Retinal prosthetics and visual cortex implants each have their respective strengths and limitations. The retinal prosthetics are easier and safer to implant, but only work if the patient has a working optic nerve, which patients with severe damage to the eyes may not have. Visual cortex implants work in a wider variety of cases, but at the cost of a more invasive surgery to implant them. Because of their respective advantages and disadvantages, researchers will continue to pursue both technologies.

Brain-to-Brain Communication

Getting images and sounds into the brain is already a major breakthrough—but DARPA has an even larger goal. It would like to see humans able to *send* sounds, images, and other kinds of information to each other through direct brain-to-brain communication. As ambitious as this goal is, current understanding of the basic workings of the brain suggests that it's possible.

In 1999, Harvard neuroscientist Garrett Stanley performed an experiment that hinted at this ability. Stanley and his team implanted electrodes in the retina of an anesthetized cat. The electrodes recorded the activity of neurons in the retina of the cat's eye as its head was turned to look at various objects. The information from the electrodes was fed into a computer. Using software Stanley's team wrote, the computer converted the neural signals into a standard video format and sent them to a screen. On the screen, you could see what the cat is looking at. The images were blurry, but clearly recognizable. When the cat's head was turned to face the branch of a tree, on the screen you could see the branch, even if you couldn't make out all the details of its bark and texture.

Stanley's research suggests that we could record experiences at the neural level and play them back later. Just as we could take a recorded video and turn it into a set of nerve impulses for your brain, we could record the actual nerve impulses that are occurring and then electrically stimulate your nerves to cause you to "perceive" the same pattern again at a later time.

In principle we could do this for all the senses—record not just what you see, but also what you hear, taste, smell, and feel, all at the level of your brain. Playing back such an experience would be a little like reliving it.

You might even be able to play that kind of sensory recording back for someone else, turning the experience you had into a set of nerve impulses that could be sent into the other person's brain, allowing him or her to experience at least the sensory parts of an event from your perspective.

Stanley's research is limited by the fact that it pulls information only out of the peripheral nervous system, not the brain itself. Stanley showed that he could plug into what a cat was seeing at the time, but not into what was going on deeper in its mind.

A research team at Vanderbilt University plans to go one step further. Under a $1 million grant from DARPA, psychology professor Jon Kaas and his colleagues are studying the possibility of communicating sounds directly from brain to brain.

In their latest experiment, they plan to insert around thirty electrodes into the auditory cortex of two rhesus monkeys. When neurons in the brain of the first monkey fire, a computer will send a signal to electrodes in the matching part of the second monkey's brain. The researchers will play sounds for the first monkey. Will the second monkey react to the sounds? Better yet, will the second monkey be able to tell what the sounds played for the first monkey are? The odds are very good that the researchers will be able to communicate something between the two. Whether the second monkey will hear an intelligible sound or something completely garbled is a harder question. Yet success with the cochlear implant suggests that it's certainly possible.

Kaas's and Stanley's research opens up other possibilities. For example, it's a general principle of the brain that, from a neural perspective, *sensation equals recollection equals imagination.* This means that, for the most part, the same neurons fire if you see something, if you remember seeing it, or if you imagine seeing it. The same is roughly true for sounds, smells, tastes, and physical sensations.

This has interesting implications. It means two humans with brain-computer interfaces could send imagined sights, sounds, and sensations back and forth to each other. It also suggests that our imaginations might be useful as input devices for computers. For example, you might imagine a picture and send it to your computer, capturing it more quickly and accurately than you could by drawing it out with your hand. The computer might not even have a video screen. It might simply project the image back into your visual cortex, giving you instant feedback and allowing you to sharpen and modify the picture you're working on inside your head.

Our understanding of the brain suggests that this kind of feed-back loop between man and machine is possible. We've already seen that we can project images, sounds, and sensations into a human brain, and both basic neuroscience and limited experiments suggest we can send them out. If humans could start collaborating with computers on the contents of their imaginations, it would revolutionize computer interfaces. Instead of having to point and click or type to instruct the computer to do something, we could simply think about it. The ability to communicate with one another directly in sights, sounds, and sensations, in addition to our current tool of language, would have just as profound an impact on human interaction.

Deep-Brain Stimulation

There are 1.5 million victims of Parkinson's disease in the United States. Parkinson's disease sufferers experience uncontrollable trembling of their limbs. The trembling starts as a minor annoyance and then grows in severity, until eventually the patient is unable to control his or her limbs. There is, as of yet, no cure for Parkinson's. Drugs like L-dopa can control the symptoms of the disease for a while, but in every patient the drugs lose effectiveness over time. At best, they delay the progression of the disease.

In some cases, Parkinson's tremors become so bad—and the drugs so ineffective—that doctors cut out part of the brain to stop them. In 1977, a French neurosurgeon named Alim-Louis Benabid was performing such an operation to remove the tremor-producing part of a patient's brain. At the outset, he electrically stimulated parts of the patient's brain to map its various functions. When he accidentally stimulated the thalamus, the Parkinson's tremors stopped. Since then, doctors have been experimenting with the technique with better and better results.

In 1997, the FDA approved deep-brain stimulators (DBSs) based on Benabid's research for the control of Parkinson's symptoms. A deep-brain stimulator is basically a single long electrode that penetrates a few inches through the patient's brain and stops at the thalamus or a nearby area. In 70 percent of patients with severe Parkinson's that no longer responds to drugs, a DBS can stop or markedly reduce symptoms. The therapy has proved so effective that the electrode has been implanted in more than thirty thousand patients worldwide. In 2002, for the first time, Medicare started paying the $30,000 to $40,000 bill for the procedure. With the cost barrier to many patients eliminated, doctors expect the therapy to rapidly grow in popularity.

Interestingly enough, in the past few years, doctors have noticed that deep-brain stimulators can affect more than Parkinson's disease and other movement disorders. In the late 1990s, researchers noticed that some patients who received deep-brain stimulators would experience short periods of depression afterward. Then they noticed the opposite: some patients implanted with DBS had a suddenly improved mood. Even some patients with lifelong depression, something they'd lived with long before Parkinson's disease, seemed to become happier and more energetic. There were also reports of patients with severe obsessive-compulsive disorder (OCD) who, after receiving a deep-brain stimulator to control their Parkinson's disease, also stopped showing the signs of the OCD.

Later investigation has shown that it's exactly where the electrodes are placed that determines whether DBS will have side effects, will depress a patient, will lift mood, or will combat obsessive-compulsive disorder. Building on these results, at least three research groups in the United States and Europe are now running human trials of deep-brain stimulation for severe untreatable depression and OCD. If those trials prove successful, we'll see the first widespread use of brain implants to control depression and OCD by the end of the decade.

Other researchers have seen additional evidence that electrical stimulation can affect emotions. In 2003, Takeshi Satow of the University of Kyoto reported that stimulating the inferior temporal gyrus of a patient undergoing surgery for epilepsy induced feelings of happiness. After five seconds of stimulation, Satow's patient smiled. With a stronger current, she burst out laughing, for no particular reason at all.

These results aren't completely unprecedented; in the 1960s, a pair of scientists working independently of each other demonstrated similar and even more powerful effects on human emotion and behavior through electrical stimulation. Their names were Robert Heath and José Delgado.

There's a famous series of pictures of Delgado taken in 1964 in Madrid that speaks volumes about his work. In the first photo, Delgado—young, dashing, and trim with a full head of black hair—stands in a bullring, and across the grass from him, still some distance away, is a large bull. In Delgado's right hand is a cape. In his left is a black box with a long antenna. In the second photo, the bull is charging—leaning forward, its head down, its hooves barely touching the ground, its horns pointed straight at Delgado's midsection.

In the last photo, Delgado has dropped the cape and pushed a button on the black box in his hand. The bull is skidding to halt, just feet in front of him. Its legs are splayed forward to arrest its charge. Dirt is flying up from its hooves.

Delgado, then doing research at Yale, had demonstrated a previously unheard-of ability to electrically affect the brain. When he pressed the button on his transmitter, it sent a signal to a receiver on the bull's head. The receiver sent an electrical pulse down electrodes implanted in the bull's hypothalamus, deep within its brain, which shut down the bull's aggression and instilled a sense of tranquillity. For as long as Delgado continued to send the signal, the bull remained calm. Even efforts by the toreadors to enrage it failed. Conversely, by pressing a different button, Delgado could stimulate a different part of the bull's limbic system and drive it instantly into a rage, regardless of what it was doing at the time.

Delgado did similar work on chimpanzees and other primates. He found that he could electrically control sleep, appetite, sexual arousal, aggression, and even social behavior. In most cases, he could change behavior at the flick of a switch. An awake and alert monkey would be asleep within seconds of Delgado's sending a current to the correct part of its brain. Its other behaviors could be altered as quickly.

In later studies, Delgado and Robert Heath, working at Tulane University, in New Orleans, independently showed that they could cause similar effects in human psychiatric patients. In a series of groundbreaking and sometimes bizarre experiments, Heath placed electrodes in the brains of some twenty-six patients suffering from severe mental illness, including incurable epilepsy, Parkinson's disease, schizophrenia, and a variety of psychiatric conditions for which the patients had been hospitalized. Heath's goal was to understand the pleasure and so-called aversive centers in the brain—regions that when stimulated cause a reaction of pleasure and distress, respectively. At first he merely recorded activity. He found that when a patient flew into a violent rage, there was a large uptick in activity in the aversive system.

Heath found that he could control those symptoms by stimulating the patient's pleasure center in the septal region of the brain. This region, part of the limbic system, has since been found to be important in a wide variety of human emotions. By stimulating this area, Heath could calm a patient in a rage, reduce anxiety, and induce smiles and giggles. Many of his patients would suddenly talk about how much they loved their doctor and how wonderful it was to be in the hospital. Heath found that he could also remove severe pain, depression, and delusions. Delgado produced similar results.

Besides Heath's and Delgado's evidence, there are other good biological reasons to believe that we can affect mood electrically. For example, most antidepressants work by increasing the amount of serotonin available in your brain. Serotonin is a neurotransmitter used by one neuron to send a signal to the next. In the brain, it produces a sense of calm and well-being. Drugs like Prozac and Zoloft prevent neurons from sucking up serotonin from the area around them. With more serotonin bouncing around, there's a better chance that some of it will bump into a receptor or another neuron. The result is a boost in mood. Another way to achieve this might be to stimulate serotonin-producing neurons to release their serotonin out into the brain. The brain's serotonin-producing neurons are concentrated in a place called the raphe nucleus. From there they send out long projections into dozens of other parts of the brain. Just one electrode attached to the raphe nucleus could trigger serotonin-producing neurons to release their serotonin, which would have a strong euphoric effect.

Brain-computer interfaces may turn out to be far more powerful and precise tools for controlling mood than drugs or gene therapy. Mind-altering drugs are rather clumsy things. Whether swallowed, injected, or smoked, drugs need to work their way through the bloodstream before they reach the brain. There, they spread indiscriminately, flooding every part of the complex nerve network. Drugs generally have an affinity for only one or a few types of neurons, but they act on *every neuron* of that type.

Drugs also take time to permeate our brains, time to work their way out of our system, and time to cause their effects. Prozac must be taken for several days before it starts to work to elevate mood. A painkiller like Vicodin operates more quickly, but still takes many minutes to an hour before it begins to have its effect. Once the effect has begun, there's no easy way to stop it. The effect will last for several hours, regardless of the desired duration.

Future neural implants, on the other hand, could stimulate one neuron while leaving the neuron next to it alone. They could alter our brains in much more precise, sophisticated ways than the simple floods of chemicals that we use today. Because they're more precise, they might be able to produce their effects with fewer side effects, and they would be under the control of the user. They would also work immediately. They might be used to reduce pain, boost mood, or instill a sense of calm with virtually no delay between activation and effect. When the mood alteration of pain suppression was no longer desired, the person wearing the neural implant could turn it off immediately, stopping the flow of current to the electrodes and ending the effect, instead of waiting for hours for a drug to be cleared from the body.

Perhaps the most intriguing possibility of neural prosthesis for managing emotions goes beyond healing: it is the potential for communicating emotions directly from brain to brain. Researchers could monitor the neural impulses that occur in the emotional centers of your brain and then stimulate a similar pattern (perhaps at a lower intensity) in someone else's brain. Just as we could record the activity in your sensory or motor areas for playback later, the activity in your emotional areas could also be recorded and replayed in the future for you or someone else.

Neuroscientists already know a reasonable amount about the brain areas involved in empathy. The amygdala and other parts of the brain are responsible for modeling the emotions of other people, giving you a sense for what they're feeling. Information from the rest of your brain is sent into the amygdala and fed into this model. With neural prosthetics, information from the emotional centers of someone else—say, a loved one—could be piped straight to your empathy center. So rather than having to guess what your spouse or child is feeling, you would simply be sensing it via the wireless link between your brains. If you wanted to sense other people's feelings less, you could choose to turn down the volume, reducing the strength of the signal that your neural interface sent into your empathy centers. Software could even decide which people's feelings to send into your empathy centers and which not to. The end result might be just like having an unusually keen sense of how others are feeling, with the option to dial that sense up or down in intensity according to whatever critera you chose.

Repairing Memory, Perception, and Cognition

In the 2001 film *Memento,* actor Guy Pierce plays Leonard, a man suffering from a kind of brain damage. Leonard is able to think and reason about the world around him. He's able to keep track of things as they're happening. But he's unable to form new long-term memories. A few minutes after something happens it's simply gone from his mind. Leonard is an extreme example of truncated memory, but such cases—and others more bizarre—do exist.

Every year around 80,000 Americans suffer a blow to the head that results in permanent brain damage of some kind. More than 700,000 people suffer strokes, and not all of them fully recover brain function. The Centers for Disease Control (CDC) in Atlanta estimates that in the United States alone, more than 5 million people are living with some sort of permanent disability resulting from injury to the brain. These people have problems that span the whole gamut of the brain's function: problems with concentration, attention, memory, and perception; lost ability to produce or comprehend language; reduced ability to plan actions or solve problems; difficulty in controlling mood and emotion; long-term personality changes; depression, fatigue, anxiety, and agitation; and many more.

Repairing these higher functions of the brain is a larger challenge than any that brain-implant researchers have tackled to date. To some degree, we've already cracked the code of vision, hearing, touch, and motion. Yet in those areas we have the advantage of something concrete to map the brain's pattern of activity to. We can compare the firing of neurons in the visual cortex to the image someone is looking at and find a relationship. We can observe neurons in the auditory cortex and see which ones fire in response to certain frequencies and volumes of sound.

The higher functions of the brain are more abstract. We don't really have a good concrete analog for concentration, or perception, or memory. And while we've made impressive progress in electrically influencing emotion, our tools are still crude at best. Learning to repair damage to the brain's higher functions is going to be a long, slow process. Yet it's also going to teach us a tremendous amount about just how the brain and mind work.

Theodore Berger at the University of Southern California, another DARPA grantee, is already working on brain implants to repair damaged long-term memory function. Researchers have known for decades that an area of the brain called the hippocampus is important in taking our current experiences and storing them away for retrieval days, months, or years later. People with damaged hippocampi have trouble forming these memories. In extreme cases, they're like Leonard, living day to day with memories of what happened before their brain damage, but nothing since then. Berger and his team hope to correct this with computer circuits that act like neurons in the hippocampus.

Berger doesn't claim to know how the hippocampus works. Instead of trying to understand it, he and his colleagues have attempted to mimic its behavior. Over the past decade, they've electrically stimulated areas of the hippocampus millions of times and observed the results. This has enabled them to build a mathematical model of this part of the brain, and a computer chip that responds to electrical signals in exactly the same way that a slice of hippocampus in a lab dish does.

For the next step, testing the chip in living animals, Berger and Sam Deadwyler, of Wake Forest University, will interface the memory chip with the damaged hippocampus of a primate. The chip itself will be located outside the monkey's skull and communicate with the primate's brain through a smaller set of electrodes implanted in the hippocampus. The goal of the study is to restore the monkey's ability to learn. If it succeeds, then Berger and his team will start to move toward trials in humans who have suffered brain damage to the hippocampus. They also believe their device may help in restoring some of the function lost in Alzheimer's disease.

Berger and Deadwyler do not stop at therapeutic uses of brain implants to restore impaired memory function but are already talking about the possibility of augmenting human memory—making it easier for us to learn new facts or recall things learned long ago. If we can understand how the brain files experiences and information away into memory, their reasoning goes, we can improve on the process. We may even be able to file away a copy of a person's memories on a computer or other outside device, enabling perfect storage of our memories for decades to come. For now, the idea of enhancing or capturing human memories is in the realm of science fiction. We're not yet close to understanding how the brain encodes memory. Yet the fact that serious researchers are contemplating it at all is provocative, and studies such as Berger's and Deadwyler's are key to developing this understanding of how memory works.

Even if the Berger and Deadwyler animal studies fail to restore memory, which is quite possible, they'll still provide the most sophisticated recordings yet of the hippocampus. They'll teach us important lessons about how memory works in the primate brain. That in turn will give researchers a better starting point for the next round of research. And if Berger's work produces any useful results at all, it will encourage other labs to begin to tackle damaged memories, attention, language abilities, and more.

Perception is another piece of the brain's repetoire of higher functions, and is the focus of MIT's James DiCarlo. DiCarlo is interested in a mental disorder called agnosia, a condition where sufferers have difficulty recognizing objects. The most famous case of agnosia is described in the title essay of Oliver Sacks's 1985 book *The Man Who Mistook His Wife for a Hat*. While confusing one's spouse with an article of clothing is uncommon, it can happen in extreme cases.

DiCarlo is looking at how the infotemporal cortex of monkeys recognizes types of objects. His research plan involves recording neural activity in this part of the brain of monkeys while they look at pictures of cats and dogs, and then at computer-generated pictures that look a little like a cat and a little like a dog. The monkeys have been trained to press one button when they see a cat and a different one when they see a dog. The trick will come when DiCarlo tries to change the monkey's perceptions. While the monkey is looking at a picture that looks more like a dog, for example, the researchers will send in a neural signal like the one produced when the monkey sees a cat. If the monkey hits the button to indicate that it's seeing a cat rather than a dog, the researchers will know they've succeeded in replacing one perception with another. This will open the way to treatments for agnosia and other disorders of perception. It will also open the possibility of rather bizarre, even psychedelic, manipulations of human perception. It would suggest, for instance, that you could look at a table and be induced to recognize it as a living thing rather than an inanimate object.

The technology for enhancing human cognition may have further practical applications. Computers are better at analyzing some kinds of data and spotting some kinds of patterns than humans. A computer can look at a series of numbers and spot a trend that a human never would. A computer can examine tens of millions of moves in a chess game and find an interesting opening. A computer that could interface with human perception might be able to overlay these insights on top of our own, giving us the benefit of its analysis with the feeling of having noticed something ourselves. We would in effect be more perceptive. At the moment this is just speculation, yet it's an interesting possibility raised by the current basic research into how the brain turns our raw sensations into our perceptions.

Some researchers are considering the possibility of using neural prostheses to repair damage to other parts of the brain. Patients with damage to Broca and Wernicke's areas of the brain have problems producing or understanding language. Patients with damage to part of the left frontal cortex can't perform arithmetic or higher mathematics. There are structural differences in the amygdala and other parts of the brain between normal and autistic children. So far no experiments have been performed in these areas of the brain, but for the first time researchers are talking seriously about the possibility of repairing this kind of brain damage through neural prosthesis. As the field matures, we'll see scientists attempting to tackle these problems. As they do, they'll make new discoveries about how the brain handles language, mathematics, and other higher functions. Those discoveries in turn will open up the possibility of interfacing with these functions, improving them, and even sharing them between individuals.

CHAPTER 10

World Wide Mind

OMEDAY IN THE future, you take the plunge and have a computer interface implanted in your brain. You're not the type to take this step lightly or quickly, but it's now been ten years since brain-computer interfaces were approved for the general public. Most of your friends have them. You watch with envy as they interact on levels that you can't. You're tired of hearing them tell you what you're missing out on. Still, you're careful. You wait for others to have the procedure. You pore over the safety statistics and most common side effects. You find the model that's least likely to be obsolete a few years down the road. And you choose a doctor who has lots of experience and a good track record in implanting the devices.

You go to the hospital in the afternoon. The anesthesiologist numbs your skull but leaves you awake. As you lie there on the operating table, the doctor makes a tiny hole in your skull, through which she inserts an incredibly light, flexible mesh of electronic circuits. Using tiny tools inserted through the same hole, she spreads the mesh across the entire surface of your brain. Once the mesh is in place, it sends millions of microscopic filaments down into your brain. Each filament is a thousandth of the width of a human hair—much smaller than even the smallest nerve cells. After this, the doctor implants a tiny transmitter and receiver just inside your skull. Finally, she replaces the small amount of bone that she had cut out and coats the area with growth factors to speed the bone's healing. Once this scar heals, there will be no visible sign of your implant at all.

After the surgery, you wait in the recovery room, feeling no different than before. The doctor visits and tells you that the procedure went well and that the interface seems to be working. You rest, eat, and go to sleep. In the morning, training begins. On the first day, the interface customizes itself to the parts of your brain that handle vision. You're shown a series of pictures on a screen. As you look at them, the implant in your brain monitors the activity of neurons in your visual cortex and records which ones fire in response to which images. After an hour, your implant has a good working map of your brain. Now, after each image flashes on the screen, you're asked to close your eyes. A picture identical to the one you saw on the screen appears in your mind. Occasionally it's wrong in some way, and you push a button to signal the discrepancy. By early afternoon, your implant is able to send you images and moving pictures with almost perfect fidelity.

You nap and wake for your second training session, which takes the form of a game. You control a spaceship on the screen. You must avoid obstacles by moving your ship up and down, right and left, forward and back, but you have no joystick or keyboard with which to do this. The computer does not respond to your voice commands. You're told to *think* about the way you want the spaceship to move. You find that by concentrating you can some-times clumsily nudge the ship one way or another. You're frus-trated, but the technician calms you down. She jokes about how long it took her to master this game and says that you're a fast learner. You relax a bit and return to the game. Control comes slowly at first, and then faster and faster as you get the hang of it. You could still do better with your hands, but you're improving.

Finally, using the skills you developed in the game, you learn how to control your implant itself. At first, you learn only how to activate a mental "screen" on which the implant places informa-tion. The screen is inside your mind, projected into your visual cortex by the electrodes of your implant. Once you learn how to activate that mental projection area, you learn how to move a cur-sor around that screen and click on the mental "buttons" the im-plant projects in your mind.

Late in the afternoon, you're discharged from the hospital. You now have all the tools necessary to explore your new abilities on your own. Your spouse, who already has an implant, takes you home. The next morning, you start another training exercise. You activate the implant, and it guides you through training it to pro-ject sound into your mind. You slip headphones over your ears and play a selection of sounds and music given to you by the hos-pital. Just as with your vision, the implant monitors your brain ac-tivity in response to the sounds you hear. By the end of the morning, you can use the implant to play sounds in your mind.

In the afternoon, the implant guides you through a set of exer-cises, training it for touch. You are instructed to touch every part of your body, systematically. You stand, sit, and lie down. You take a hot bath and a cold shower. Through this, the implant learns the specifics of how the sense of touch affects your brain. By doing so, it learns how to stimulate your brain to mimic these effects.

The next day, you start the most intensive phase of training the implant: learning to communicate with language. You relax as the implant uses sight and sound to present you with words, phrases, and sentences. You repeat these to yourself. The implant shows you images, which you name. It instructs you to read paragraphs and pages off your mental screen. It teaches you to press a mental "send" button at the same time you mentally speak. By the end of the first day, you have the rudimentary ability to "type" words to the implant by thinking about them. The next day, you practice again. By the end of the third day, you can dictate to the implant as quickly as you can think.

On the following day, you learn to communicate with more than words. The implant shows you how to contact other people with implants. You practice sending words back and forth with your spouse. You're stunned by the ability to communicate silently. Then the implant shows you how to send other senses. You send your spouse what you see and hear and feel. You close your eyes and imagine a flower and a romantic tune and send those as well. Your spouse responds in kind. You begin to explore this incredible way to communicate. You find that you can project emotions and even some abstract ideas. In one conversation, you flit back and forth among words, images, sounds, and feelings. That night, you and your spouse make love while opening all of your senses and emotions to each other. The intimacy is beyond anything you have known. It's almost overwhelming, and you have to retreat. But you know that you will try this again, and that your relationship will reach new depths of intimacy.

The next day, you return to work. Over the coming weeks, you find your work productivity greatly increased. You can work faster, more easily, and more accurately with teammates who have implants than you can with the nonimplanted. In meetings with other implanted workers, you routinely work using mental diagrams, images, and silent speech. In mixed company, you take mental control of the presentation, projecting text and images onto the screen and changing them with a facility you never would have reached with a marker and whiteboard.

In the evenings, you continue to explore your implant. You learn how to use it to control sleep, appetite, mood, sexual arousal, and even your heart rate and respiration. You gain the ability to remain calm merely by willing it. You learn how to use the implant to navigate the Internet, asking questions and receiving answers just by thinking about them. The implant guides you through increasingly complex exercises. You play memory games with numbers, shapes, colors, names, faces, places, and all sorts of objects. You add, subtract, and multiply. You play strategy games and solve mental puzzles. Through all of these activities, the implant studies your brain's activity, learning to interface with your highest functions.

Little by little, you feel yourself getting sharper, as the implant adds to your mental abilities. You phrase mathematical questions to yourself and the implant supplies the answers instantly. You find yourself able to remember twenty digits at a time, where before you could only remember seven. Your attention wanders less at work and you're able to juggle more things in your mind simultaneously. The answers to questions you pose to the Internet come back more quickly and sometimes almost before you ask them. You routinely trade memories and experiences with other implanted humans. You learn to view the world through other people's eyes. You let others see through yours. As the months and years pass, you increasingly view your implant as a vital and natural part of you. Using it becomes as natural as breathing. You can no longer imagine a disconnected life.

Science Fiction to Reality

This scenario is science fiction, yet everything in it is grounded in the research discussed in the previous chapter. What has been discovered about the brain so far suggests that in the coming decades, we'll be able to link our minds to computers in deep ways, and through those computers to each other.

At the same time, it's important to separate the incredible potential of brain-computer interfaces from the here and now. Blindness and paralysis will not be cured in 2006 or 2008 or even 2010. We won't be mentally transmitting our thoughts to each other in 2015. It took twenty years for the cochlear implant to move from the first experimental successes to Food and Drug Administration approval for mainstream use. It took just as long—from 1977 to 1997—for the results of research into electrically suppressing Parkinson's tremors to make it into a device approved for mass use.

There are two key hurdles that brain-computer interfaces need to clear before they'll be widely employed therapeutically—to combat blindness, paralysis, memory loss, and the other problems they've been shown to be effective against—let alone adopted by the healthy masses as a new communication tool. The first hurdle is limitations on information flow because of the low number of interfacing neurons. Current systems in use in human beings can interact with anywhere from one neuron, in Johnny Ray's case, to 256 neurons, in the case of the visual cortex prosthesis that's partially restored Jens Naumann's vision. It's amazing that these systems have produced useful results at all, considering that both vision and motor control involve billions of neurons.

To go from experimental proof of concept to real applications, future implants will need to connect to many more neurons. Naumann sees in a 16-by-16-pixel grid. By way of comparison, a normal television broadcast is a grid 480 by 440—more than 200,000 pixels. A good computer screen has more than a million pixels. To provide Naumann with something closer to normal human vision, researchers will need to connect to more neurons at a time.

Fortunately, that's quite doable. Neural interfaces have such low bandwidth because they're built on decades-old technology. For most of the field's history, it's been populated by neurosurgeons and neurobiologists—scientists who understand the brain, but aren't on the cutting edge of electronics. Over the last few years, researchers with backgrounds in computer science and electrical engineering have entered the field of brain-computer interfaces. They've brought with them the latest expertise in fabricating circuits and a goal of greatly increasing the number of connections implants make with the brain.

Researcher Ken Wise at the University of Michigan, for example, is producing implants fabricated like microchips. Wise's lab is just finishing its 1,024-channel wireless implant. It measures just 3 millimeters on a side and communicates wirelessly through the skull at 1.5 megabits per second. The German firm Infineon is working on a neurochip that can connect to 16,384 neurons at a time.

These devices are a major improvement on current systems, but they still have plenty of room to improve. Intel's Pentium 4 processor is not much larger than existing neurochips and has 55 million transistors on it. It has about forty times as many circuits as the optic nerve has neurons, in about the same cross-sectional area. Circuits built by the latest chip-fabrication techniques are actually much smaller and more tightly packed than neurons in the brain; most of the basic technology used to build Pentium chips can be applied to building better, higher-bandwidth brain implants. As the field of neural prosthetics matures, it will borrow more and more of these techniques from the computer industry, enabling the creation of brain implants that can connect to millions of neurons at a time. Eventually the limitation will not be the hardware, but the number of neurons that we can physically connect to without damaging the brain.

The second—and more significant—hurdle is brain surgery. Cutting open the skull and placing something inside the brain incurs sizable risks. For example, one out of every fifty patients to receive a deep-brain stimulator suffers some kind of bleeding in the brain afterward. In rare cases, this can prove fatal. Even putting aside the risk, brain surgery is an expensive process that often takes days to recover from. Deep-brain stimulators cost an average $60,000 to install. Cochlear implants cost only slightly less. Many brain surgeries, such as aneurysm repair, require three or more days of recovery in a hospital before the patient can go home. The combination of these issues will give anyone considering brain surgery pause. If neural interfaces are ever to become mainstream, whether for their medical use or as enhancements, scientists and physicians will have to dramatically lower the health risk, cost, and inconvenience of tapping into the brain.

Researchers are looking at two ways to get around these problems. The first possibility is to develop entirely noninvasive interfaces. Instead of actually attaching electrodes to neurons, future brain-computer interfaces might be able to monitor and interact with individual neurons from outside the skull, without a need for surgery at all. DARPA in particular is pouring money into this avenue of research.

Today there are two top technologies that could deliver a noninvasive interface. The first is electroencephalography, or EEG, a technique for recording and interpreting the electrical activity of the brain. An EEG device works by means of one to more than one hundred electrodes placed on the scalp, held in place by sticky pads. EEG devices pick up the aggregate electrical activity of the 100 billion neurons in the brain. EEG scientists talk about brain activity in terms of "brain waves." Neurons, it turns out, tend to fire with a certain rhythm. Exactly how fast they fire depends on what kind of neurons they are, where in your brain they are, and what you're doing with your mind at the time. The more relaxed you are, the slower your neurons are firing. In deepest sleep, most of your brain is active at around one to four firings per second, or 1 to 4 hertz. As you become more active, different neurons in the different parts of your brain that are being put to use start to fire more quickly. EEG devices frequently see 30-hertz activity in various parts of the brain, and sometimes faster.

Researchers Niels Birbaumer and Jonathan Wolpaw have used EEGs as a kind of brain-computer interface for paralyzed patients. They've painstakingly trained quadriplegics to change their brain-wave activity. The EEG picks up the change and interprets it as a set of commands. Birbaumer and Wolpaw's patients have used this technique to control powered wheelchairs and have moved cursors around on computer screens, much like Johnny Ray. EEG devices are fairly inexpensive to build and are small enough to carry around with you. If manufactured in volume, they might cost as little as a few hundred dollars each.

Unfortunately, Birbaumer and Wolpaw's research seems to have pushed EEG about as far as it can go. To be picked up by an EEG device, any electrical signal from a neuron has to pass through the skull and scalp (and usually many other neurons) before reaching an electrode; thus EEGs produce at best very general information about what's going on inside the brain. With EEG it's impossible to get the kind of precise information about specific neurons and patterns of neural firing that you can get from inside the brain. For that reason, most brain-computer interface researchers think EEG won't go much further than it's gone already.

The other noninvasive technology is functional magnetic resonance imaging, or fMRI. fMRI is the most advanced technique for looking at the brain now widely in use. An fMRI machine can detect changes in blood flow in areas as small as one millimeter on a side. When neurons fire, they draw more blood to replenish their supply of nutrients and oxygen. By detecting blood flow, fMRI detects which parts of the brain are most active. fMRI has three main drawbacks. First, even one millimeter on a side isn't very high resolution when dealing with the brain. In a one-millimeter cube, there may be twenty thousand neurons. Second, fMRI machines generate huge magnetic fields, and they need magnets that weigh several tons to generate those fields. The machines cannot be used anywhere near metal objects. Finally, fMRI is expensive, up to several million dollars per device.

Despite the limitations of EEG and fMRI, we shouldn't rule out the possibility of a noninvasive interface. Researchers are pursuing many different avenues, including picking up the magnetic fields of neurons, shining infrared lasers through the outermost layers of the brain to pick up changes in opacity caused by neural firing, and even more exotic methods. In early 2003, scientists used an advanced microscope that fires light into its target to take pictures of single neurons. The technology is fairly bulky and can see only a tiny area at a time, but future generations will likely shrink and increase in capacity. Researchers have also used something called transcranial magnetic stimulation (TMS) to change the activity of neurons in the brain. By applying a magnetic field just outside the skull, they can temporarily scramble the electrical signals in a small part of the brain. After the magnetic field is gone, brain activity in the affected area returns to normal. Scientists use this as a safe way to probe the workings of the brain, by effectively turning off a small area and seeing how a subject is able to function. The spatial resolution of TMS falls somewhere between that of EEG and of fMRI. At this time, it can be used only with whole areas of the brain, not targeted to individual neurons. Its low precision means that it can be used only to scramble brain activity, not to induce specific patterns of activity. Thus its use as a brain-computer interface is limited, but researchers are working on ways to improve its precision.

Interfacing with individual neurons without opening up the skull is a long shot today, but one that's generated quite a bit of interest, including general medical interest. High-resolution non-invasive scans of the brain would prove invaluable as diagnostic tools, giving doctors the ability to spot tumors and burst blood vessels and watch brain activity in real time. That's why hundreds of millions of dollars are going into research projects aimed at improving the resolution of noninvasive scanning and manipulation technologies.

Another line of research is to improve the surgical implantation process itself to the point where it's simple enough, inexpensive enough, and safe enough to be performed routinely. There's precedent for surgical techniques advancing to the point where millions of people have them on a purely elective basis. For example, consider plastic surgery. Liposuction and breast augmentation are major surgeries and are performed hundreds of thousands of times each year.

A better analogy may be corrective eye surgery. As of 2002, more than 10 million people worldwide have had corrective eye surgery. The vast majority of these people already had virtually perfect vision by way of corrective lenses, so for them, corrective eye surgery was purely a matter of convenience or vanity. The only benefit it brings is freedom from having to wear glasses or contact lenses.

As recently as 1990, the idea of surgically correcting vision was exotic, even a bit bizarre. The most sophisticated corrective eye surgery of the time, radial keratotomy (RK), involved making several incisions in the eye. It carried high risks and a recovery time of up to a week for each eye. That year, twenty thousand corrective eye surgeries were performed worldwide. What changed corrective eye surgery was the excimer laser; the LASIK procedure, which utilizes the excimer laser, can be done in about fifteen minutes. Recovery can be as quick as a single day; the risk of complications is far lower than with RK.

Is there an equivalent of the excimer laser and LASIK for brain surgery? It's possible. A leading contender is endovascular ("in-vein") surgery, in which no incision is made in the skull at all. Instead, doctors insert a probe through a hole in an artery, usually in the thigh, and guide it through the circulatory system to the brain. Using magnets placed around the head and real-time brain scans, physicians can then steer the device to the exact point in the brain they want to reach.

Endovascular surgery in the brain was first attempted in the early 1990s and was approved by the FDA for use in the United States in 1995. One common use is delivering doses of radiation or chemotherapy directly to the site of a tumor without having to affect the rest of the brain. The most frequent use, though, is in the treatment of aneurysms.

An aneurysm is a weak area in a blood vessel of the brain. The pressure of the blood at the weak point causes the vessel to balloon out, stretching it further. Eventually it bursts, causing bleeding in the brain, sometimes severe enough to be categorized as a stroke.

Today physicians estimate that a staggering 18 million people are walking around with unruptured aneurysms. Prior to endovascular neurosurgery, repairing an aneurysm (either before or after it burst) was a several-hour procedure that involved drilling through the skull. It typically required the patient to remain in the hospital for three to four days after the surgery. Endovascular surgery, on the other hand, can repair aneurysms in as little as ninety minutes, and the patient is frequently able to leave the hospital on the same day. Follow-ups a year later show that patients who had aneurysms repaired through endovascular procedures recovered more quickly and had fewer complications than patients who went through traditional brain surgery. As a result, the procedure is becoming more and more common, with a total of about 125,000 endovascular treatments of aneurysms performed between 1995 and 2002.

In 2002, Rudolpho Llinas, one of the elder statesmen of brain science, proposed a way to use the principles of endovascular surgery to reach millions of neurons at once. Llinas, chairman of the New York University Department of Neuroscience and the editor in chief of the journal *Neuroscience,* helped pioneer the study of how neurons function. Llinas's idea—which he is exploring under a grant from the National Science Foundation—is to use an existing network of passageways that reaches every neuron in the brain—the network of arteries, veins, and capillaries that supply blood to these neurons—to create a neurovascular brain-computer interface. A probe would be inserted into an artery in the body and guided up to the brain via magnets outside the body, much as it would be in any endovascular procedure. The probe is actually a collection of a million or so much smaller wires, each roughly a tenth of a micrometer thick. As the probe is threaded through the brain's vascular networks, the flow of blood and gentle guidance from the magnets separate the wires from each other, allowing them to be carried into a million arterial nooks and crannies. The smallest capillaries in the brain are about ten micrometers wide, or one fifth the diameter of a human hair. The wires Llinas proposes to use are one hundred times thinner. Scientists have already created wires that thin—and even thinner—using carbon nanotubes.

The advantage of Llinas's proposal is that one wouldn't need to drill through the skull, or even penetrate the blood-brain barrier (a double-thick wall of cells that prevents blood from touching nerve cells) to insert the interface. Nor would doctors have to push an electrode through layers of neurons to reach the ones deeper inside the brain—the network of blood vessels extends everywhere. What's more, the system would be completely removable. It could simply be pulled back out the same way it went in.

Llinas and his team are currently doing proof-of-concept tests in animals using just a few wires at a time. If the proposal proves viable, receiving a future brain-computer interface may be as simple as a ninety-minute outpatient procedure, at which point it would become more a more acceptable option for both medical and enhancement uses.

Safe, easy to install, multimillion-neuron brain implants would revolutionize the field of neural prosthetics. They would allow us to restore lost vision and hearing to the levels enjoyed by healthy humans. Paralyzed patients would be able to control robotic limbs and computer interfaces with the same ease and precision with which the healthy control their natural limbs.

Higher-bandwidth implants would likely give us more subtle, fine-grained control over our emotions than the comparatively crude experiments of Delgado and Heath. They may allow us to project images, sounds, and sensations as rich as we can imagine to our computers and to each other, and even allow us to communicate our inner emotions to each other. Most important, these massively connected implants would provide new insights into memory, attention, language, perception, and the other higher functions of the brain. And such understanding may in turn lead to implants that *enhance* our existing mental functions.

Nonzero Minds

Neural interfaces bring together the two great information-processing systems that exist today: the brain and the computer. The brain's power manifests itself in the flexibility of the human mind. We have language, politics, science, and culture. We excel in learning new skills and adapting to new situations. We're general problem solvers, able to take lessons learned in one domain and apply them to another.

By contrast, our computers are simple, rigid, specialized things. An e-mail program will never learn to handle voice-mail, despite the similarity between the two. Deep Blue, the chess program that beat the world chess champion, Gary Kasparov, can't tell you what it thinks the crucial move was. It speaks only in chess moves. It can beat the world champion chess player, but it doesn't know how to play checkers. In fact, a three-year-old human can do countless things this specialized computing chess champion can't.

Nevertheless, our computers can do things that our brains cannot. While humans can do complex, general things well, computers specialize in doing simple things at incredible speeds. They can calculate, sort, search, filter, and transform simple information millions of times faster than humans, without ever becoming bored, tired, or inattentive. They can store and access vast amounts of data and communicate it at phenomenal speeds, without ever understanding what it means.

The capabilities of brains and computers are complementary— that's why we build computers. The more tightly we can couple those computers to the information processing that happens in our minds, the more we'll benefit. The evidence suggests that we could couple these new abilities and our old ones quite closely. Our brains already have an evolved facility with tools that lends itself to the incorporation of computers as parts of our self-image. The last few decades of research into neuroscience have shown that the brain reorganizes itself in response to reading, to playing the piano, or to other interactions with the tools and devices we've built. In an interview he gave in 2004 after completing one of his most recent studies of a primate's use of a robot arm, Miguel Nicolelis said:

Our analyses of the brain signals showed that the animal learned to assimilate the robot arm into her brain as if it was her own arm. . . . Such findings tell us that the brain is so amazingly adaptable that it can incorporate an external device into its own neuronal space as a natural extension of the body. Actually, we see this every day, when we use any tool, from a pencil to a car. As we learn to use that tool, we incorporate the properties of that tool into our brain, which makes us proficient in using it.

Phil Kennedy observed the same thing in his patient Johnny Ray, who was the first human to control a computer through a neural interface. Although Ray struggled at first with the interface, within a few months he could manipulate the cursor naturally, without even thinking. Six months after the surgery, when Kennedy asked what it felt like to move the computer cursor, Ray spelled out "N-O-T-H-I-N-G."

That level of integration suggests that neural prostheses that affect our higher cognitive abilities will also feel like extensions of ourselves, rather than external devices. If so, we'll be able to marry our strengths of learning, creativity, and especially very *general* thinking and problem solving with the computer strengths of amazing speed, precision, storage capacity, and very *specific* and narrow information processing. We'll be able to take advantage of silicon speed, accuracy, and storage in the domains that our minds are uniquely suited to—problem solving, language, invention, art, culture, and emotion. We'll achieve something of the best of both worlds.

Fundamentally, neural interfaces are a new information technology. The history of information technologies shows that they frequently have their largest impact through their effects on communication. Neural interfaces will give humans the ability to communicate in an integrated stream of language, images, sounds, sensations, and emotions. We'll combine these different modalities just as fluidly as we combine speech and hand gestures today, but to much greater effect. One friend telling another about a great date may tell a story in words, images of her handsome escort for the night, the rich sound of the symphony playing, the taste of the wine they drank, the feel of his hand on hers, and the emotional rush of their goodnight kiss.

Those new communication abilities, like our current repertoire of abilities, could work as well with groups as with one-on-one pairs. When sensations, emotions, and ideas become digital, it's as easy to share them with a dozen friends, or a thousand strangers, as it is to send them to one person. Just as we can e-mail our words, place them on websites, or send them in instant messages or chat rooms, we'll be able to broadcast the inner states of our minds. When and with whom we choose to share such things will be up to us. The technology will give us the flexibility to do what we like with the contents of our thoughts, feelings, and imaginations, and society will respond with new social norms to guide our choices.

Mind-to-mind communication will find applications in the workplace as well. Imagine two engineers planning a project together. As they talk, they project mental images to one another—crude schematics of the structure they plan to build. They produce their mental images much more quickly than they could draw them with pencil and paper or mouse and keyboard. Computer software mediates their communication, capturing the images, allowing both of them to work on editing the plan at once, and storing their work for future reference or for publication to others.

As the engineers mentally sketch, the computer assists their work. It runs calculations on strength and weight, supplying the information as necessary. It rotates the image more fluidly than a human mind can, letting the engineers see the structure from all sides. It stores a more detailed image than humans can hold in their minds at once. As each engineer focuses on part of the structure, embellishing and clarifying and refining, the computer integrates it with the whole. In effect, it increases the power of their visual imagination, just as would a drafting board, but far more quickly, intimately, and intuitively.

Scientists will also profit from the new interfaces. Albert Einstein once said that he owed his discovery of relativity to his ability to visualize space and time. Future Einsteins may discuss their work with their colleagues as much in imagery as in words. The idea of a "thought experiment" becomes more powerful when thoughts can be projected to others, or assisted by software and hardware outside the mind.

There are similar implications for the arts and the communications media. Once we have direct connections between our minds, the concept "multimedia" will take on a whole new dimension. We'll see immersive "movies" that touch all of our senses, our emotions, and more. Direct neural connections to the sensory portions of our brains would allow us to build virtual worlds with one another that are as far beyond the crude virtual-reality goggles available today as movies are beyond comic books. Mature neural prostheses would affect every type of mass communication, from movies to poetry, from music videos to educational texts, from live concerts to museums. All would be enriched by the ability to communicate in images, sounds, sensations, and emotions.

The closest parallel in history may be the event that ushered in the Renaissance—the development of the printing press by Johannes Gutenberg in the 1450s. Prior to Gutenberg, books were rare objects, limited to the extremely wealthy. To spread the information stored in a book, a scribe had to manually copy it into another bound volume. Human copiers kept the cost of books high while also introducing errors into each successive copy. The printing press changed that. For the first time, humanity had an efficient way to precisely share large amounts of information. Men and women who never met—who were not even alive at the same time—could share their thoughts with each other in a structured and durable fashion.

In addition, the printing press ushered in a surprising decentralization of society. Prior to Gutenberg, the most powerful entity in Europe was the Catholic church. The first work that Gutenberg printed on his press was the Bible. Yet Martin Luther used the printing press to publish his pamphlets attacking the Church, thus spurring the Protestant Reformation. The press did help the Church spread its own message, but it had a larger effect in magnifying the voices of those who were not yet being heard.

Over the course of history, information and communications technology—the printing press, phones, computers, and the Internet—have generally empowered individuals and undermined dictatorial states. They've facilitated change rather than control. Photocopying technology and fax machines ushered in reform in the former Soviet Union, and Internet access is slowly facilitating change in China today (despite attempts at censorship by the state). In contrast to centrally controlled information technologies such as television and radio, newer technologies like the World Wide Web, cell phones, and instant messaging are *distributed*. Any person can share original or passed-on information with any other person or group. Neural interfaces would be the pinnacle of distributed communication devices, allowing each individual to transmit her thoughts, senses, emotions, and knowledge to any similarly equipped person in the world. History suggests that this development will empower individuals and groups, and undermine central authority.

While undermining central authority, information technology nevertheless increases the benefits of human cooperation. In his book *Nonzero: The Logic of Human Destiny*, Robert Wright proposes a fascinating theory of human history that explains this trend. Wright draws on a mathematical field known as game theory. In game theory, there are "zero-sum" games, in which players can benefit only at the expense of another player. There are also "non-zero-sum" games, where all players can benefit without needing to harm the others in order to do so. Looking back over twenty thousand years of human history, Wright comes to the conclusion that humankind has repeatedly developed new ways to communicate and coordinate our activities to make our interactions more non-zero-sum. That is to say, we've invented new ways of working together so that we all benefit.

The printing press is a perfect example of this. An author who puts down his work in a book loses very little, and may indeed gain for himself. At the same time, the entire rest of the world may gain tremendously by consuming the knowledge he shares. The total gain for the world far exceeds the loss or effort put in.

The net effect of information technology is an increase in global intelligence. We individuals are, in a sense, like neurons in a global brain—a marketplace of ideas, experiences, and innovations. The more power we gain to communicate with one another, the more integrated that aggregate brain becomes. In the last few centuries, we've taken tremendous steps, from small isolated pockets of computation in individual tribes and civilizations to the World Wide Web where anyone with a computer can publish text, images, sounds, and videos to a billion others around the world. The next step is the integration of our biological brains: unlocking the inner ideas and experiences we have, and allowing us to share them with one another, to weave them together into thoughts in a world wide mind.

CHAPTER 11

Life Without Limits

LTAMIRA CAVE SITS just inland of the northern coast of Spain, five kilometers from the scenic town of Santillana del Mar. Green fields and gently rolling hills dot the countryside. The historic buildings and cobblestone streets of the town date back to the sixteenth century.

In 1868, a hunter following his dog stumbled across the cave. For the next decade, other hunters used it as a refuge from the elements. Yet its importance would be discovered by a young girl. Altamira sits on what was then the estate of Don Marcelino Sanz de Sautuola, a Spanish nobleman and amateur archaeologist. In 1879, Marcelino brought his nine-year-old daughter, Maria, with him on an exploration of the cave. While he looked down at the floor of the cave searching for bones and arrowheads, it was she who pointed up at the ceiling and exclaimed, *"Mira, Papa, bueyes!"* (Look, Papa, oxen!). Marcelino and his daughter had discovered the first cave paintings left by Stone Age people.

For almost another twenty years, archaeologists would maintain that the paintings at Altamira were an elaborate hoax. Yet by the turn of the century (sadly, after Marcelino's death), even the staunchest critics were convinced that the vivid paintings of bison and other animals in the cave had been painted at least ten thousand years ago, well before recorded history.

The discovery of Altamira has been followed by dozens of other finds of cave paintings, burial grounds, jewelry, sculpture, and sophisticated tools dating back tens of thousands of years. Archaeologists and anthropologists interpret these artifacts as the first signs of true culture, the turning point that marks the people who left them as fully modern humans, possibly even as the first indisputable sign of human language.

This flourishing of culture occurred around forty thousand years ago. Prior to that time, the archaeological record contains primitive general-purpose stone cutting tools, signs that wooden tools were made as well, and clear evidence of the use of fire. The skeletons look fully anatomically human, and the skulls are large enough to accommodate modern-sized brains, but there is no art, there are no multipart tools, there are no signs of burial of the dead, nor is there anything else that would separate the humans of this time culturally from those who lived tens of thousands of years earlier.

Suddenly, however, the archaeological record changes. Stone tools become more specialized: instead of simple sharpened rocks, there are awls, picks, needles, and other shapes. Bone fishhooks appear, as do harpoons, spear throwers, and soon thereafter bows and arrows. And for the first time people were making cave paintings like those at Altamira.

The people who left those artifacts forty thousand years ago were identical to us in skeletal structure and probably in DNA as well. Human evolution at a biological level hasn't moved much from that point. But human cultural evolution has moved at an ever-accelerating rate. Within a few tens of thousands of years of the first cave paintings, our ancestors had domesticated horses and other animals, had discovered agriculture, had built permanent dwellings for themselves where none had been before, and were on their way to discovering metallurgy.

A few short thousands of years after that, they developed writing and mathematics, astronomy and the wheel, philosophy and the beginnings of medicine. Since then, progress has continued to accelerate. In just the last century, we've unraveled the mysteries of gravity, electricity, and light. We've charted much of the visible universe. We've built machines that can manipulate matter and energy to move us from place to place, to shelter us from the elements, to make distant objects seem close, to send our words and images to one another, to store and manipulate huge amounts of information, to perform acts of physical labor that would have been the work of entire nations in the not-so-distant past. Through increasingly sophisticated tools, we have already enhanced ourselves.

In the past two centuries, we've learned as much about the world inside us as the world around us. We've discovered that our bodies are composed of cells, which themselves are made up of proteins, which are described by genes. We've learned that our brains and nervous systems are made up of special kinds of cells—neurons—and we have learned how they communicate with one another. That knowledge has allowed us to heal the sick and is on the verge of allowing us to enhance our minds and bodies.

Every century, every decade, every year things seem to change more quickly. It took tens of thousands of years to go from cave paintings to agriculture. It took thousands of years to go from agriculture to writing. It took thousands more to go from writing to the first steam engine. Yet it took less than seventy years to go from the first automobiles to humans walking on the moon. It was just twenty years between Jim Watson and Francis Crick's discovery of the structure of DNA to Cohen and Boyer's first success at genetic engineering, and less than thirty more to the complete mapping of the human genome.

Now that same accelerating pace of change is on the verge of touching us in the most intimate way—by giving us the power to reshape our own minds and bodies.

There are those who want to put the brakes on this kind of change. Environmentalist Bill McKibben is one of them. If we change our minds and bodies as we've changed the world around us, he fears, we'll lose sight of the very meaning of our lives. In his book *Enough: Staying Human in an Engineered Age,* he writes, "We are snipping the very last weight holding us to the ground, and when it's gone we will float silently away into the vacuum of meaninglessness."

McKibben's fear is that we are defined by our limitations. If we exceed them, we will no longer be human. That's why he wants us to say "Enough." Yet throughout our history, we've exceeded our limits and added to our capabilities. If our limits define us, then we stopped being human a long time ago, when we invented tools and language and science that extended the powers of our minds and bodies beyond those our hunter-gatherer ancestors were born with.

Leon Kass seems to see things more clearly. In a passage urging respect for our given state, he writes: "The human soul yearns for, longs for, aspires to some condition, some state, some goal toward which our earthly activities are directed but which cannot be attained in earthly life. Our soul's reach exceeds our grasp; it seeks more than continuance; it reaches for something beyond us, something that for the most part eludes us."

History suggests that Kass is correct. We have always sought to reach beyond our grasp. This continual yearning for more that he fears is in fact a hallmark of humanity. It has driven the evolution of our culture since the Great Leap Forward forty thousand years ago, and continues to drive our progress today.

Only in mythical utopias or dystopias do we ever see a permanent sense of contentment. In the real world, contentment is transitory. It comes with each accomplishment. If it lasted too long, it would stunt our urge to grow and change.

This hunger, this reach that exceeds our grasp, this aspiration to attain something "which cannot be attained in earthly life" is the force that has built our world. It has built our comfortable lives. It has produced the medicine that keeps us alive. It has assembled our vast store of knowledge. It has produced our art, our music, our philosophy. It has built our deepest understanding of the mysteries of the universe. Never to say enough, always to want more—that is what it means to be human.

We owe everything about our current lives to our predecessors on this earth, the ones who would not stand for "enough" and instead asked, "What next?" Those brave, foolhardy inventors and explorers sought out better ways to live, ways to live in more comfort, and with greater health, and with greater opportunities for their children. Their curiosity, their willingness to experiment with the unknown, their bravery in the face of risk has brought us everything—from agriculture to written language to the use of fire to telephones to antibiotics and more. The debt we owe them can never be repaid—they're gone from this world, beyond the reach of our gratitude.

If we can never repay that debt, we can perhaps pay it forward to future generations. We cannot perfectly predict, may not even be able to imagine, the ways that our descendants will put the technologies we develop to use. Yet the lesson of history is that they will use the powers we provide them to make their world a better place. Now it's time to do what past generations did for us, to explore the world, to experiment with new ways of doing and being, so that we can pass what we learn on to the future—not to try to dictate the lives of future generations, but to let countless families and individuals chart their own course, apply their own best judgments, and use their collective intelligence to propel our species into the future.

Destiny of Species

The chain of events that has led us to this point stretches back into the deep history of our world. Six million years ago, our prehuman ancestors diverged from the evolutionary line that would eventually produce our chimpanzee relatives. It wasn't a sharp divergence—we didn't grow large brains or start composing sonnets overnight. Another 3 million years passed before we started to use the first primitive stone tools, and more millennia passed before we mastered language and fire. Yet viewed from the long perspective of the history of life on earth, the development from our chimp ancestor until today has happened with extraordinary speed.

In those 6 million years, we've come to stand upright. We've shed our body hair. We've evolved to live longer lives. Most impressively, our brains have ballooned in size, and with that expansion has come the capacity for tools, speech, writing, and the trappings of modern society, which have so recently appeared.

Brain Size of Human Ancestors

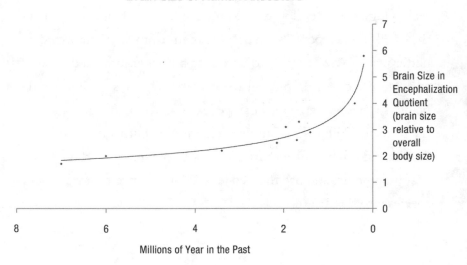

FIGURE 10 Human brain size adjusted for body weight has tripled in the last 7 million years, with the trend accelerating over time.

To place this in context, life on earth existed for almost 4 billion years before the first organism with a brain appeared. Sixty million years later, our chimp-ancestor's brains was a prodigious 400 cubic centimeters. In the 6 million years since then, the human brain has evolved to a volume of 1,350 cubic centimeters on average, a tripling in size. Something remarkable is going on.

Evolution does not, in general, have a direction. It has no grand plan. It is, in the words of philosopher Daniel Dennett, an algorithm—a simple rule mechanically applied over and over again. If a gene is effective at passing itself to future generations, it will flourish. If a gene is not effective at reaching future generations, it will die out.

Sometimes that process leads to more complex and more intelligent creatures. Other times it leads to simplification, as evolution selects for the most efficient members and species shed traits they no longer need. For the most part, evolution leads in neither direction. The dominant form of life on earth, in terms of number of individuals, number of species, and sheer biomass, are single-celled bacteria. From an ecological point of view, this is the age of bacteria, as it has been for 3.5 billion years.

While the evolutionary process itself has no direction, the details of a particular environment or species may impose one, at least for a period of time. In a volcanic crater, any life that evolves will be resistant to heat. In Antarctica, evolution will select for resistance to cold. A species that gives birth to only a few children will evolve to invest heavily in each child, while a species that lays thousands of eggs will evolve to invest less in each offspring.

Some combination of such factors—both environmental and biological—came together in prehuman history to propel the evolution of one splinter off the primate branch of the tree of life toward large brains, sophisticated tools, rich communication abilities, and long lives. A number of explanations for this amazing sequence of events have been put forward. Evolutionary psychologists propose that we exist in a "cognitive niche," a particular area in the evolutionary landscape that we've been progressively selected to better fit. University of Washington neuroscientist William Calvin has proposed that frequent and severe climate changes over the last 3 million years selected for creatures with big brains who could easily adapt to new conditions. Biologist George Miller, in his book *The Mating Mind,* theorizes that sexual selection—the choice of whom to mate with—rather than natural selection of the survival sort has driven the development of large brains in humans.

We may never know for certain what triggered this evolutionary surge. What we do know is that it signaled a phase shift—a marked change in the tempo—in our evolution. Suddenly, 6 million years ago, change in one branch of the tree of life accelerated and new complexity exploded onto the scene. This was not the first such transition in evolutionary history, nor will it be the last.

Life on earth began almost 4 billion years ago. For the first 3 billion of those years, all organisms were single-celled. Only about 700 million years ago did multicelled creatures evolve, and with their presence on earth came a huge flowering of the varieties of life. The Cambrian explosion, which started around 540 million years ago, produced all the basic body plans of creatures found on earth today in just a few tens of millions of years. For some reason, after almost 3 billion years in which evolution produced little more complexity than algae or bacteria, everything changed in the blink of an eye. New complexity and new diversity appeared. A new branch of life on earth came into being, one that represented a paradigm change, from single cells and simple colonies, to complex creatures that could swim, walk, and fly and even crudely contemplate the world around them.

We humans represent the next such phase shift. Our arrival on this world is as significant in biological terms as that of the first multicelled creatures. We are as different from every past life-form on this planet as chimpanzees are from bacteria. We alone possess the power to alter our own minds and bodies and those of our children. We alone possess the power to guide our own development—to choose our paths, rather than allowing nature to blindly select for the genes that are best at spreading themselves.

We are not the end point of evolution—there is no such thing.

We are just an intermediate step on one branch of the tree of life. But from this point on, we can choose the directions in which we grow and change. We can choose new states that benefit us and benefit our children, rather than benefiting our genes.

We will not all opt for the same changes. We will not all choose the same direction of travel. Different men and women, different communities, different ideologies will all select different goals to work toward. Some of us will choose to stay as we are, while others will choose to transform. Humanity will expand, splinter, and blossom.

At some point, one hundred years or one thousand years or one million years from now, our world and perhaps this corner of our universe will be populated by descendants whom we might not recognize. Yet they will think, and love, and dream of better tomorrows, and strive to achieve them, just like us. They will have the traits most dear to us. They will be different in ways we cannot imagine.

Most of all, these descendants of ours will be fantastically diverse. Where we are all alike in the basic forms of our minds and bodies, they will exist in a plethora of forms. Humanity will have given birth not to one new species, not to a dozen new species, but to thousands, or millions. We will have spawned a new explosion of life as sudden and momentous as that of the Cambrian explosion. Long after we are gone, after there no longer live any beings with our DNA, our distant descendants will still look back. They'll look at this moment in our history and marvel. *To think that such primitive creatures as* Homo sapiens *could give birth to whole new kingdoms of life! They could not have understood what they were creating! Yet we're lucky that they had such a strong urge to change and grow—that is why we're here.*

We are, if we choose to be, the seed from which wondrous new kinds of life can grow. We are the prospective parents of new and unimaginable creatures. We are the tiny metazoan from which a new Cambrian can spring. I can think of no more beautiful destiny for any species, no more privileged place in history, than to be the initiators of this new genesis.

Acknowledgments

This book wouldn't exist without the efforts of a large number of people who helped me cross the gulf between the idea and the realization. Especially for a first-time author, the encouragement, mentorship, and support I received from friends and strangers alike was invaluable.

At the very beginning, Leo Parker and Todd Kopriva gave me much-needed encouragement and constructive criticism of my writing. Without them I doubt I would have gotten past the first chapter.

A number of other authors, some of whom I knew and others whom I approached out of the blue, took the time to give me advice on writing a book or to critique what I'd written already. I'm especially grateful to Scotto Moore, Dan Farmer, Erik Davis, Mark Pesce, Charles Mann, Greg Stock, Robert Wright, and David Brin.

I count myself lucky to have a large circle of friends and associates who were willing to read early drafts or excerpts and provide their critical feedback. Among them are Adam Henderson, Alexis Carlson, Anders Sandberg, Ania Mitros, Antoun Nabhan, Aubrey de Grey, Brent Field, Brett Paatsch, Chris Phoenix, Daniel Westreich, David Lockhart, David Perlman, Elizabeth Kister, Emily Rocke, Gabriel Williams, James Hughes, John Rae-Grant, Lesley Carmichael, Lori Kingery, Natalia Powers, Marcelo Calbucci, Mike Clark,

Nancy Linford, Parag Mallick, Rafal Smigdrozki, Raven Hanna, Robert Bradbury, Ryan Grant, Sarah Black, Simon Windor, Stuart Updegrave, and Ty Landrum.

Finally, I must thank my research assistant, Tess Bruney, who helped ensure that my claims rested on sound scientific ground, my agent, Ted Weinstein, whose sage advice guided me through every step of the process, and my editor, Becky Cole, whose critical eye and incredible enthusiasm helped make this the book it is.

Notes

Introduction: Healing and Enhancing

3 Leon Kass: For a good analysis of Kass's history, see Chris Mooney, "Irrationalist in Chief," *The American Prospect,* September 24, 2001.

4 "there are certain things": Francis Fukuyama, *Our Posthuman Future: Consequences of the Biotechnology Revolution* (New York: Farrar, Straus and Giroux, 2002).

4 "A ban on cloning": William Kristol, *The Future Is Now: America Confronts the New Genetics* (Lanham, Md.: Rowman & Littlefield, 2002).

4 "We need to do an unlikely thing": Bill McKibben, *Enough: Staying Human in an Engineered Age* (New York: Henry Holt, 2003).

6 "Regulation that flies in the face of reality": Steven E. Hyman, M.D., "Pediatric Psychopharmacology," address to the Presidential Council on Bioethics, session 1, Arlington, Virginia, March 6, 2003, transcript available at http://www.bioethics.gov/transcripts/march03/session1.html.

7 "The finitude of life is a blessing": Leon Kass, *Life, Liberty, and the Defense of Dignity: The Challenge of Bioethics* (San Francisco: Encounter Books, 2002).

Chapter 1: Choosing Our Bodies

11 Ashanti DeSilva's therapy: For the story of Ashanti DeSilva and the first gene therapy attempt on a human being, see W. French Anderson, "Gene Therapy: The Best of Times, the Worst of Times," http://www.frenchanderson.org/french/docarticles/best.html; National Institutes of Health, "Results from First Human Gene Therapy Clinical Trial," October 19, 1995, http://www.hhs.gov/news/press/1995pres/951019a.html; R. Blaese et al., "T Lymphocyte-Directed Gene Therapy for ADA-SCID: Initial Trial Results After 4 Years," *Science* 270 (October 20, 1995): 475–80; Claudio Bordignon et al., "Gene Therapy in Peripheral Blood Lymphocytes and Bone Marrow for ADA-Immunodeficient Patients," *Science* 270 (October 20, 1995): 470–75.

13 Cohen and Boyer: For biographies of Stanley Cohen and Herbert Boyer, the winners of the 1996 Lemelson-MIT Prize, see http://web.mit.edu/invent/a-winners/a-boyercohen.html.

15 "What choice did we have?": Ruth Sorelle, "The Gene Doctors," *Houston Chronicle,* http://www.chron.com/content/houston/interactive/special/gene/gfirst.html.

16 gene therapy blossomed: Tim Friend, "Elusive Gene Therapy Forges On," *USA Today,* February 23, 2003, http://www.usatoday.com/news/health/2003-02-23-gene-therapy_x.htm.

17 Anemia patient statistics: See Leslie Hanna, "Drug Watch: Epoietin Alfa (EPO) for Anemia," in Project Inform, *The HIV Drug Book,* 2nd ed. (New York: Pocket Books, 1998).

17 kidneys produce EPO: J. W. Adamson, "The Promise of Recombinant Human Erythropoietin," review, *Seminars in Hematology* 26, Supplement 2 (April 1989): 5–8; S. K. Tripathy et al., "Stable Delivery of Physiologic Levels of Recombinant Erythropoietin to the Systemic Circulation by Intramuscular Injection of Replication-Defective Adenovirus," *Proceedings of the National Academy of Science of the USA* 91 (November 1994): 11557–61; Jerry E. Bishop, "Gene Therapy Method May Offer Hope in Treating Anemia with Protein EPO," *Wall Street Journal,* November 10, 1993.

17 The cost of EPO therapy: See John Tagliabue, "Mystery Effect in Biotech Drug Puts Its Maker on Defensive," *New York Times,* October 2, 2002.

17 first animal study of EPO gene therapy: Jeffrey M. Leiden et al., "Long-Term Erythropoietin Expression in Rodents and Non-human Primates Following Intramuscular Injection of a Replication Defective Adenoviral Vector," *Human Gene Therapy* 8 (October 10, 1997): 1797.

18 monkeys went from: See S. Zhou et al., "Adeno-Associated Virus-Mediated Delivery of Erythropoietin Leads to Sustained Elevation of Hematocrit in Nonhuman Primates," *Gene Therapy* 5, no. 5 (May 1998): 665–70.

18 Chiron and Ariad: Victor M. Rivera et al., "Long-Term Regulated Gene Expression in Non Human Primates," *Molecular Therapy* 7 (May 2003).

20 Team Festina: B. J. Spalding, "Black-Market Biotechnology: Athletes Abuse EPO and HGH," *Biotechnology* 9, no. 11 (November 1991): 1050, 1052–53; Christie Aschwanden, "Gene Cheats," *The New Scientist,* January 15, 2000; "Drug Slush Fund: Festina's Doctor Says Team Manager Obtained Drugs for Riders," CNN/*Sports Illustrated* website, July 21, 1998, http://sports illustrated.cnn.com/cycling/1998/tourdefrance/news/1998/07/21/ drugs_slush/.

20 EPO can boost your performance: Glenn Zorpette, "All Doped Up—and Going for the Gold," *Scientific American* (May 21, 2000): 20, 22.

21 Other athletes have used: E. Randy Eichner, "Ergogenic Aids: What Athletes Are Using—and Why," *The Physician and Sports Medicine* 25, no. 4 (April 4, 1997), available at http://www.physsportsmed.com/issues/1997/ 04apr/eichner.htm.

23 Katie Binley: K. Binley et al., "Long-Term Reversal of Chronic Anemia Using a Hypoxia-Regulated Erythropoietin Gene Therapy," *Blood* 100, no. 7 (October 1, 2002): 2406–13; R. Pollock and T. Clackson, "Dimerizer-Regulated Gene Expression," *Current Opinion in Biotechnology* 13, no. 5 (October 1, 2002): 459–67.

24 ALS: For an overview of ALS, see the website of the Oregon and Southwest Washington Chapter of the ALS Association, http://www.alsa-or. org/info/Statistics.

25 IGF-1 gene therapy study: Results are reported in B. K. Kaspar et al., "Retrograde Viral Delivery of IGF-1 Prolongs Survival in a Mouse ALS Model," *Science* 301 (August 8, 2003): 839–42.

25 IGF-1 gene therapy to boost mouse muscle strength: See E. R. Barton-Davis et al., "Viral Mediated Expression of Insulin-like Growth Factor I

Blocks the Aging-Related Loss of Skeletal Muscle Function," *Proceedings of the National Academy of Sciences of the USA* 95 (December 1998): 15603–7; H. Lee Sweeney, "Enhancement 5: Genetic Enhancement of Muscle," address to President's Council on Bioethics (PCBE), Washington, D.C., September 13, 2002, transcript available at http://www.bioethics.gov/transcripts/sep02/session7.html; A. Musaro et al., "Localized IGF-1 Transgene Expression Sustains Hypertrophy and Regeneration in Senescent Skeletal Muscle," *Nature Genetics* 27, no. 2 (February 2001): 195–200.

26   Geoffrey Goldspink's work: See M. Hill and G. Goldspink, "Expression and Splicing of the Insulin-like Growth Factor Gene in Rodent Muscle Is Associated with Muscle Satellite (Stem) Cell Activation Following Local Tissue Damage," part 2, *Journal of Physiology* 549 (June 1, 2003): 409–18.

26   Se-Jin Lee's work: See Alexandra C. McPherron, Ann M. Lawler, and Se-Jin Lee, "Regulation of Skeletal Muscle Mass in Mice by a New TGF-ß Superfamily Member," *Nature* 387 (May 1, 1997): 83–90.

27   knocking out myostatin: S. Bogdanovich et al., "Functional Improvement of Dystrophic Muscle by Myostatin Blockade," *Nature* 420 (November 28, 2002): 418–21; McPherron, Lawler, and Lee, "Regulation of Skeletal Muscle Mass."

27   sarcopenia: W. Evans and I. Rosenberg, *Biomarkers* (New York: Simon & Schuster, 1991); "Sarcopenia: Origins and Clinical Relevance," review, *Journal of Nutrition* 127, Suppl. 5 (May 1997): 990S–91S.

27   "Just say you'd like your pectoralis muscle": Sweeney, "Genetic Enhancement of Muscle."

28   more than 8 million people: PlasticSurgeryResearch.com, "Cosmetic Plastic Surgery Research: Statistics and Trends for 2001, 2002, and 2003," http://www.comesticplasticsurgerystatistics.com/statistics.html.

28   leptin gene therapy: Patrick Muzzin et al., "Correction of Obesity and Diabetes in Genetically Obese Mice by Leptin Gene Therapy," *Proceedings of the National Academy of Science of the USA* 93 (December 1996): 14804–8.

28   "couch potato's dream": Eric Benjamin Lowe, "UF Researchers Explore Gene Therapy to Treat Obesity," *Friday Evening Post* (University of Florida

Health Science Online Newsletter), June 25, 1999, http://www.health.
ufl.edu/post/jun99.html.

29 EpiTan: See EpiTan company home page for information on its research
and products, http://www.epitan.com.au/.

29 albino mice whose skin can change color: C. A. Cronin, W. Gluba, and
H. Scrable, "The Lac Operator-Repressor System Is Functional in the
Mouse," *Genes and Development* 15 (June 15, 2001): 1506–17.

29 gene therapy for baldness: Norimitsu Saito et al., "High Efficiency Genetic
Modification of Hair Follicles and Growing Hair Shafts," *Proceedings of the
National Academy of Science of the USA* 99, no. 20 (October 1, 2002):
13120–24.

31 French leukemia cases: For information on these cases, see S. Hacein-Bey-
Abina et al., "A Serious Adverse Event After Successful Gene Therapy for
X-Linked Severe Combined Immunodeficiency," *New England Journal of
Medicine* 348, no. 3 (January 16, 2003): 255–56; Jo Lyford, "Gene Therapy
'Caused T-Cell Leukemia,' " *The Scientist,* October 20, 2003; http://www.
biomedcentral.com/news/20031020/02, accessed on November 1, 2003; A
Fischer and M. Cavazzana-Calvo, "*LMO2*-Associated Clonal T-Cell Prolifer-
ation in Two Patients After Gene Therapy for SCID-X1," *Science* 302 (Octo-
ber 17, 2003): 415–19.

34 DNA nanoball gene therapy: Sylvia Pagán Westphal, "DNA Nanoballs
Boost Gene Therapy," *The New Scientist,* May 12, 2002.

35 intron gene therapy: See H. Guo et al., "Group II Introns Designed to Insert into Therapeutically Relevant DNA Target Sites in Human Cells," *Science* 289 (July 21, 2000): 452–57.

35 "more like a rifle shot": University of Texas at Austin, "Cellular Mechanisms for Delivering Genetic Therapies," press release, July 28, 2000, available at http://www.newswise.com/articles/2000/7/INTRON.TXA.html.

36 "bidding defiance": Andrew Dickson White, *A History of the Warfare of Science with Theology in Christendom* (New York: Appleton, 1898).

36 Queen Victoria: Donald Caton, "John Snow's Practice of Obstetric Anes-
thesia," *Anesthesiology* 92 (2000): 247–52.

37 one tenth of 1 percent: Tufts University, Center for the Study of Drug Development, "Center for the Study of Drug Development Pegs Cost of a

New Prescription Medicine at $802 Million," press release, November 30, 2001, available at http://csdd.tufts.edu/NewsEvents/RecentNews.asp?newsid=6.

38 first artificial-heart patient: "First Artificial Heart Recipient, Robert Tools, Dies at 59," Associated Press, November 30, 2001.

38 "What's the worst that can happen to me?": Ricki Lewis, *Human Genetics: Concepts and Applications* (New York: McGraw-Hill Science), 392.

39 $17 billion: *Entrepreneur,* January 2001.

39 emergency-room visits: DanceSafe, "The Need for Adulterant Screening Programs," http://www.dancesafe.org/documents/druginfo/pilltesting.php.

40 unsafe abortion: World Health Organization, "Abortion: A Tabulation of Available Data on the Frequency and Mortality of Unsafe Abortion," report, 3rd ed. (Geneva, Switzerland: World Health Organization, in press).

Chapter 2: Choosing Our Minds

42 statistics on Alzheimer's disease: Alzheimer's Association website, "About Alzheimer's—Statistics," http://www.alz.org/AboutAD/Statistics.asp.

42 estimated National Institutes of Health spending: Aubrey de Grey, personal communication, New Haven, Connecticut, September 2003.

43 NGF gene therapy: M. H. Tuszynski et al., "Nerve Growth Factor Gene Therapy for Alzheimer's Disease," *Journal of Molecular Neuroscience* 19, nos. 1 and 2 (August–October 2002): 207.

43 NGF gene therapy human trials: For a review of these findings, see M. H. Tuszynski, "Growth-Factor Gene Therapy for Neurodegenerative Disorders," *The Lancet Neurology* 1, no. 1 (May 2002): 51–57; for information on ongoing clinical trials, see M. H. Tuszynski, "Gene Therapy for Alzheimer's Disease: Clinical Trial Information," available at Tuszynski Lab website, http://tuszynskilab.ucsd.edu/.clinical_study.htm, posted December 20, 2002.

44 *improve* the learning and memory of normal mice: A. I. Brooks, D. A. Cory-Slechta, and H. J. Federoff, "Gene-Experience Interaction Alters the Cholinergic Septohippocampal Pathway of Mice," *Proceedings of the*

National Academy of Science of the USA 97, no. 24 (November 21, 2000): 13378–83.

44 Doogie mice: J. Z. Tsien, "Building a Brainier Mouse," *Scientific American* (April 2000): 62–68; Y. P. Tang et al., "Genetic Enhancement of Learning and Memory in Mice," *Nature* 401 (September 2, 1999): 63–69.

46 feel major pain longer: F. Wei et al., "Genetic Enhancement of Inflammatory Pain by Forebrain NR2B Overexpression," *Nature Neuroscience* 4, no. 2 (February 2001): 164–69; D. L. Stull, "Better Mouse Memory Comes at a Price," *The Scientist* 15, no. 7 (April 2, 2001): 21.

47 Alzheimer's patients had lower-than-normal activity: C. A. Dickey et al., "Selectively Reduced Expression of Synaptic Plasticity-Related Genes in Amyloid Precursor Protein and Presenilin-1 Transgenic Mice," *Journal of Neuroscience* 23, no. 12 (June 15, 2003): 5219–26.

47 CREB: Nicholas Wade, "3 Share Nobel Prize in Medicine for Studies of the Brain," *New York Times,* October 10, 2000; B. K. Kaang, E. R. Kandel, and S. G. Grant, "Activation of cAMP-Responsive Genes by Stimuli That Produce Long-Term Facilitation in Aplysia Sensory Neurons," *Neuronology* 10, no. 3 (March 1993): 427–35; P. K. Dash, B. Hochner, and E. R. Kandel, "Injection of the cAMP-Responsive Element into the Nucleus of Aplysia Sensory Neurons Blocks Long-Term Facilitation," *Nature* 345 (June 21, 2002): 718–21.

47 fruit flies had astoundingly good memories: J. Yin and Tim Tully, "CREB as a Memory Modulator: Induced Expression of a dCREB2 Activator Isoform Enhances Long-Term Memory in *Drosophila,*" *Cell* 81 (1995): 107–15.

48 could learn things in two trials: M. Barad et al., "Rolipram, a Type IV-Specific Phosphodiesterase Inhibitor, Facilitates the Establishment of Long-Lasting Long-Term Potentiation and Improves Memory," *Proceedings of the National Academy of Science of the USA* 95, no. 25 (December 8, 1998): 15020–25; Tim Tully et al., "Targeting the CREB Pathway for Memory Enhancers," *Nature Reviews Drug Discovery* 2 (April 2003): 267.

48 Memory Pharmaceuticals and Helicon Therapeutics: More information about these companies is available at their websites: http://www.memorypharma.com and http://www.helicontherapeutics.com, respectively.

48 prevalence of Alzheimer's: The large number of people suffering from

Alzheimer's disease or age-associated memory impairment and its impact on the race for memory-boosting drugs is well documented: J. J. Sramek, A. E. Veroff, and N. R. Cutler, "The Status of Ongoing Trials for Mild Cognitive Impairment," *Expert Opinion Investigating Drugs* 10, no. 4 (April 2001): 741–52; L. Helmuth, "All in Your Mind," *SAGE KE—Science of Aging Knowledge Environment* (an online "knowledge environment" resource for researchers in the field of aging) 8 (February 26, 2003); Robert Langreth, "Viagra for the Brain," *Forbes*, February 4, 2002, available at http://www.forbes.com/forbes/2002/0204/046.html; President's Council on Bioethics, "'Better' Memories? The Promise and Perils of Pharmacological Interventions," staff working paper, http://bioethicsprint.bioethics.gov/background/better_memories.html.

49   "Memory enhancers could become 'lifestyle' drugs": Tully et al., "Targeting the CREB Pathway for Memory Enhancers," 267.

49   "It's not an 'if'—it's a 'when' ": Langreth, "Viagra for the Brain."

50   caffeine and nicotine: M. S. Mumenthaler et al., "Psychoactive Drugs and Pilot Performance: A Comparison of Nicotine, Donepezil, and Alcohol Effects," *Neuropsychopharmacology* 28, no. 7 (July 2003): 1366–73.

50   15 million prescriptions: Testimony of DEA Deputy Director Terrence Woodworth to Congress, May 16, 2000, available at http://www.dea.gov/pubs/cngrtest/ct051600.htm.

51   same mental boosts to normal children and adults: Judith Rapoport, "Dextroamphetamine: Its Cognitive and Behavioral Effects in Normal and Hyperactive Boys and Normal Men," *Archives of General Psychiatry* 37, no. 8 (August 27, 1980): 933–43.

51   "Even if you have never been diagnosed": Ken Livingston, "Ritalin: Miracle Drug or Cop-Out," *The Public Interest* (Spring 1997): 3–18.

51   "universal performance enhancer": Diller's comments to President's Council on Bioethics.

51   modafinil and Provigil: For a discussion of these drugs, see Amanda Onion, "The No-Doze Soldier: Military Seeking Radical Ways of Stumping Need for Sleep," ABCnews.com, December 18, 2002, http://abcnews.go.com/sections/scitech/DailyNews/nosleep021218.html; A. Buguet, D. E. Moroz, and M. W. Radomski, "Modafinil—Medical Considerations for Use in Sustained Operations," *Aviation, Space, and Environmental Medi-*

cine 74, no. 6 (June 2003): 659–63; R. Pigeau et al., "Modafinil, D-amphetamine and Placebo During 64 Hours of Sustained Mental Work. I. Effects on Mood, Fatigue, Cognitive Performance and Body Temperature," *Journal of Sleep Research* 4, no. 4 (December 1995): 212–28; D. M. Batejat and D. P. Lagarde, "Naps and Modafinil as Countermeasures for the Effects of Sleep Deprivation on Cognitive Performance," *Aviation, Space, and Environmental Medicine* 70, no. 5 (May 1999): 493–98; T. J. Lyons and J. French, "Modafinil: The Unique Properties of a New Stimulant," *Aviation, Space, and Environmental Medicine,* 62, no. 5 (May 1991): 432–35; A. Buguet et al., "Modafinil, D-amphetamine and Placebo During 64 Hours of Sustained Mental Work. II. Effects on Two Nights of Recovery Sleep," *Journal of Sleep Research* 4, no. 4 (December 1995): 229–41.

52   effects of antidepressants on nondepressed humans: See B. Knutson et al., "Selective Alteration of Personality and Social Behavior by Serotonergic Intervention," *American Journal of Psychiatry* 155 (1998): 373–79; R. J. Verkes et al., "Reduction by Paroxetine of Suicidal Behavior in Patients with Repeated Suicide Attempts but Not Major Depression," *American Journal of Psychiatry* 155 (1998): 543–47.

52   link between serotonin and social dominance in primates: See M. J. Raleigh et al., "Dominant Social Status Facilitates the Behavioral Effects of Serotonergic Agonists," *Brain Research* 348, no. 2 (December 2, 1985): 274–82; M. J. Raleigh, G. L. Brammer, and M. T. McGuire, "Male Dominance, Serotonergic Systems, and the Behavioral and Physiological Effects of Drugs in Vervet Monkeys (*Cercopithecus aethiops sabaeus*)," *Progress in Clinical Biological Research* 131 (1983): 185–97.

53   genetic contributions to ADHD: Research is summarized in the Online Mendelian Inheritance in Man (OMIM) database maintained by Johns Hopkins University, http://www.ncbi.nlm.nih.gov/entrez/query.fcgi?CMD=Search&DB=omim, and search for "Attention Deficit-Hyperactivity Disorder," entry 143465.

53   bipolar-disorder genes: E. Hattori et al., "Polymorphisms at the G72/G30 Gene Locus, on 13q33, Are Associated with Bipolar Disorder in Two Independent Pedigree Series," *American Journal of Human Genetics* 72, no. 5 (May 2003): 1131–40.

53   mutations in a gene responsible for the serotonin transporters: K. P. Lesch

et al., "Primary Structure of the Serotonin Transporter in Unipolar Depression and Bipolar Disorder," *Biological Psychiatry* 37, no. 4 (February 15, 1995): 215–23.

53 gene therapy for pain control: For a summary of the research, see D. Fink, M. Mata, and J. C. Glorioso, "Cell and Gene Therapy in the Treatment of Pain," *Advanced Drug Delivery Reviews* 55, no. 8 (August 28, 2003): 1055–64; University of Pittsburgh Medical Center, "Cancer Pain Control Possible with Gene Therapy," press release, October 15, 2002.

54 cut their drinking habits: P. K. Thanos et al., "Overexpression of Dopamine D2 Receptors Reduces Alcohol Self-Administration," *Journal of Neurochemistry* 78, no. 5 (September 2001): 1094–1103.

54 prairie voles: L. J. Pitkow et al., "Facilitation of Affiliation and Pair-Bond Formation by Vasopressin Receptor Gene Transfer into the Ventral Forebrain of a Monogamous Vole," *Journal of Neuroscience* 21, no. 18 (September 15, 2001): 7392–96.

55 5-HT2-C receptor and psychedelic and spiritual experiences: See J. LeDoux, "The Self: Clues from the Brain," *Annals of the New York Academy of Science* 1001 (October 2003): 295–304; E. C. Lauterbach, A. Abdelhamid, and J. B. Annandale, "Posthallucinogen-like Visual Illusions (Palinopsia) with Risperidone in a Patient Without Previous Hallucinogen Exposure: Possible Relation to Serotonin 5HT2a Receptor Blockade," *Pharmacopsychiatry* 33, no. 1 (January 2000): 38–41; R. K. Kurup and P. A. Kurup, "Hypothalamic Digoxin, Hemispheric Chemical Dominance, and Spirituality," *International Journal of Neuropsychopharmacology* 113, no. 3 (March 2003): 383–93.

55 genetic link to homosexuality: It is clear from population genetics that there is some linkage, but so far it is not clear whether the chromosome region Dean Hamer identified contributes to it. See D. H. Hamer et al., "A Linkage Between DNA Markers on the X Chromosome and Male Sexual Orientation," *Science* 261 (1993): 321–27.

55 genetic linkage to religious experience: See T. J. Bouchard, Jr., et al., "Intrinsic and Extrinsic Religiousness: Genetic and Environmental Influences and Personality Correlates," *Twin Research* 2, no. 2 (June 1999): 88–98; Karen Carver and J. Richard Udry, "The Biosocial Transmission of Religious Attitudes," presentation at The National Longitudinal Study of Ado-

lescent Health conference, 1997, abstract available online at http://www.
cpc.unc.edu/projects/addhealth/abs2.html.

56   $300 billion per year: James H. Scully, Jr., testimony before the Pennsylvania House Insurance and House Health and Human Services committees on the Mental Health Parity Act House Bill 2237, PN 1872, August 15, 1996.

56   reducing the incidence of Alzheimer's: Citizens for Long Term Care, "Long Term Care: An Overview," http://citizensforltc.org/issues_overview.html.

57   average IQ and rate of economic growth: Richard Lynn and Tatu Vanhanen, "Intelligence and the Wealth and Poverty of Nations," 2002, online at Richard Lynn's website, http://www.rlynn.co.uk/pages/articles.htm.

57   every additional year of education: Association of University Teachers, "Losing Out—How UK Higher Education Is Losing Vital Investment," report, June 2000, available at http://www.aut.org.uk/media/html/losing vitalinvestment1.html.

58   attitudes toward genetic engineering: Johns Hopkins Genetics and Public Policy Center and Princeton Survey Research Associates, "Public Awareness and Attitudes about Reproductive Genetic Technology," report, December 9, 2002, available at http://www.dnapolicy.org/research/ reproductiveGenetics.jhtml?subSection=reproductive.

Chapter 3: Created Equal

63   close to a billion dollars: Tufts University, Center for the Study of Drug Development, "Tufts Center for the Study of Drug Development Pegs Cost of a New Prescription Medicine at $802 Million," press release, November 30, 2001, available at http://csdd.tufts.edu/NewsEvents/RecentNews.asp? newsid=6.

63   drugs whose patents have expired: U.S. Congress, Congressional Budget Office, "How Increased Competition from Generic Drugs Has Affected Prices and Returns in the Pharmaceutical Industry," July 1998, available at http://www.cbo.gov/showdoc.cfm?index=655&sequence=0.

64   Figure 1: U.S. Congress, Congressional Budget Office, "How Increased Competition from Generic Drugs Has Affected Prices and Returns in the

Pharmaceutical Industry," July 1998, available at http://www.cbo.gov/showdoc.cfm?index=655&sequence=0.

67 life expectancy and GDP: Data courtesy of World Resources Institute and the *UN Human Development Report,* 2003.

68 Figure 2: Courtesy of World Resources Institute and the *UN Human Development Report,* 2003.

76 $6,911 per student: AdApted from Frank Johnson, "Revenues and Expenditures for Public Elementary and Secondary Education: School Year 1999–2000," report (Washington, D.C.: U.S. Department of Education, National Center for Education Statistics, May 16, 2002).

Chapter 4: Methuselah's Genes

78 In 1900, an American male: L. Hayflick, "How and Why We Age," review, *Experimental Gerontology* 33, nos. 7–8 (November–December 1998): 639–53.

79 Figure 4: L. Hayflick, "How and Why We Age," review, *Experimental Gerontology* 33, nos. 7–8 (November–December 1998): 639–53.

80 "If, indeed, we were to . . .": S. Jay Olshansky, "Duration of Life: Is There a Biological Warranty Period?" testimony before President's Council on Bioethics, December 12, 2002, available at http://www.bioethics.gov/transcripts/dec02/session2.html.

81 $3.6 billion: "Antiaging Products," *Entrepreneur* (January 2001).

82 "Position Statement on Human Aging": S. Jay Olshansky, Leonard Hayflick, and Bruce A. Carnes, "The Truth About Human Aging," *Scientific American* (May 2002): 78–81.

82 "in the past few years": Gordon Lithgow, address to Anti-Aging Drug Discovery Summit, San Francisco, October 2002.

82 Andrzej Bartke: Bartke et al., "Extending the Lifespan of Long-Lived Mice," *Nature* 414 (November 22, 2001): 412.

83 change to a single gene could double the life span: T. E. Johnson, "Increased Life Span of Age-1 Mutants in *Caenorhabditis elegans* and Lower Gompertz Rate of Aging," *Science* 249 (1990): 908–12.

84 "Worms have 100 million nucleotides": Steve Austad, testimony to President's Council on Bioethics, December 12, 2002, http://www.bioethics.gov/december02/session1.html.

84 "My initial findings": Tom Johnson, personal communication (e-mail), June 2004.

84 genes that can increase life span: A comprehensive database of genetic changes that lengthen animal life span can be found at *Science*'s *SAGE KE—Science of Aging Knowledge Environment,* http://sageke.sciencemag.org.

85 more resistant to stresses of all sorts: G. J. Lithgow et al., "Thermotolerance and Extended Life-Span Conferred by Single-Gene Mutations and Induced by Thermal Stress," *Proceedings of the National Academy of Science of the USA* 92, no. 16 (August 1, 1995): 7540–44.

86 added resistance to stresses: Toren Finkel and Nikki J. Holbrook, "Oxidants, Oxidative Stress and the Biology of Ageing," *Nature* 408 (November 9, 2000): 239–47.

87 how they work in mice: Huber R. Warner, "Subfield History: Use of Model Organisms in the Search for Human Aging Genes," *Science*'s *SAGE KE—Science of Aging Knowledge Environment* (February 12, 2003).

88 Holzenberger's mice: M. Holzenberger et al., "IGF-1 Receptor Regulates Lifespan and Resistance to Oxidative Stress in Mice," *Nature* 421 (December 4, 2002): 182–87.

88 Researchers at Harvard: M. Bluher, B. B. Kahn, and R. C. Kahn, "Extended Longevity in Mice Lacking the Insulin Receptor in Adipose Tissue," *Science* 299 (January 24, 2003): 572–74.

90 "wonderful": R. J. Davenport, "One for All," *Science*'s *SAGE KE—Science of Aging Knowledge Environment* (December 11, 2002), http://sageke.sciencemag.org/cgi/content/full/sageke;2002/49/nf15.

93 free-radical theory of aging: For reviews of this theory, see H. K. Biesalski, "Free Radical Theory of Aging," review, *Current Opinion in Clinical Nutrition and Metabolic Care* 5, no. 1 (January 2002): 5–10, and A. Kowald, "The Mitochondrial Theory of Aging," review, *Biological Signals and Receptors* 10, nos. 3–4 (May–August 2001): 162–75.

94 produce more SOD: J. Sun et al., "Induced Overexpression of Mitochondrial Mn-superoxide Dismutase Extends the Life Span of Adult *Drosophila melanogaster,*" *Genetics* 161, no. 2 (June 2002): 661–72.

94 methuselah mice: C. C. Spencer et al., "Testing an 'Aging Gene' in Long-Lived *Drosophila* Strains: Increased Longevity Depends on Sex and Genetic Background," *Aging Cell* 2, no. 2 (April 2003): 123–30.

94   p66shc seems to make the cells: E. Migliaccio et al., "The p66shc Adaptor Protein Controls Oxidative Stress Response and Life Span in Mammals," *Nature* 402 (November 18, 1999): 309–13.

94   EUK-8 and EUK-134: S. Melov et al., "Extension of Life-Span with Superoxide Dismutase/Catalase Mimetics," *Science* 289 (September 1, 2000): 1567–69.

Chapter 5: Choosing Our Life Spans

96   caloric restriction: R. Weindruch and R. L. Walford, *The Retardation of Aging and Disease by Dietary Restriction* (Springfield, Ill.: Charles C Thomas, 1988).

97   known for almost a century: S. D. Hursting et al., "Calorie Restriction, Aging, and Cancer Prevention: Mechanisms of Action and Applicability to Humans," *Annual Review of Medicine* 54 (2003): 131–52.

98   "One of the striking things": Steve Austad, testimony to President's Council on Bioethics, December 12, 2002, http://www.bioethics.gov/december02/session1.html.

99   slows the rate of death of neurons: University of Florida, "UF Study: Calorie Restriction Reduces Age-Related Brain Cell Death," press release, December 30, 2002, available at http://www.napa.ufl.edu/2002news/localoriebrains.htm.

99   biochemistry of younger animals: G. S. Roth et al., "Biomarkers of Caloric Restriction May Predict Longevity in Humans," *Science* 297 (August 2, 2002): 811.

99   compressed morbidity: E. J. Masoro, "Subfield History: Caloric Restriction, Slowing Aging, and Extending Life," *Science's SAGE KE—Science of Aging Knowledge Environment* 8 (February 26, 2003).

100 effects of caloric restriction on rhesus monkeys: G. S. Roth, D. K. Ingram, and M. A. Lane, "Caloric Restriction in Primates and Relevance to Humans," *Annals of the New York Academy of Science* 928 (April 2001): 305–15.

101 side effects of caloric restriction: K. Hopkin, "Dietary Drawbacks," *Science's SAGE KE—Science of Aging Knowledge Environment* 8 (February 26, 2003).

102 INDY gene: Blanka Rogina et al., "Extended Lifespan Conferred by Cotrans-porter Gene Mutations in Drosophila," *Science* 290 (December 15, 2000): 2137–40.

103 Rpd3 lived longer: Blanka Rogina, Stephen L. Helfand, and Stewart Frankel, "Longevity Regulation by *Drosophila* Rpd3 Deacetylase and Caloric Restriction," *Science* 298 (November 29, 2002): 1745.

103 4-phenyl butyrate extended their lives: Hyung-Lyun Kang, Seymour Ben-zer, and Kyung-Tai Min, "Life Extension in *Drosophila* by Feeding a Drug," *Proceedings of the National Academy of Science of the USA* 99, no. 2 (2002): 838–43.

103 caloric restriction mimetics: M. A. Lane, D. K. Ingram, and G. S. Roth, "The Serious Search for an Anti-Aging pill," *Scientific American* (August 2002): 36–41.

104 genes that are expressed differently: The technology developed by Bio-marker Pharmaceuticals, Inc., is explained at their website on the page "Technology," http://www.biomarkerinc.com/html/technology.htm.

105 all turn out to be related: G. Gat-Yablonski et al., "Leptin Reverses the In-hibitory Effect of Caloric Restriction on Longitudinal Growth," *Endocrinol-ogy* 154, no. 1 (January 2004): 343–50.

105 preserve healthy, youthful life: C. K. Lee et al., "Gene Expression Profile of Aging and Its Retardation by Caloric Restriction," *Science* 285 (August 27, 1999): 1390–93; R. S. Sohal and R. Weindruch, "Oxidative Stress, Caloric Restriction, and Aging," *Science* 273 (July 5, 1996): 59–63.

106 live around 75 percent longer: M. Beckman, "Dieting Dwarves Live It Up," *Science's SAGE KE—Science of Aging Knowledge Environment,* 2001.

106 Figure 5: M. Beckhan, "Dieting Dwarves Live It Up," *Science's SAGE KE—Science of Aging Knowledge Environment,* 2001.

107 "Indefinite postponement of aging may be within sight": A. D. de Grey et al., "Time to Talk SENS: Critiquing the Immutability of Human Aging," *Annals of the New York Academy of Science* 959 (April 2002): 452–62, 463–65 (discussion).

109 succeeded in moving a mitochondrial gene: D. P. Gearing, P. Nagley, and J. Embo, "Yeast Mitochondrial ATPase Subunit 8, Normally a Mitochondr-ial Gene Product, Expressed in Vitro and Imported Back into the Or-ganelle," *EMBO Journal* 5, no. 13 (December 20, 1986): 3651–55.

111 "The lights are green everywhere you go": Gina Kolata, "Pushing Limits of the Human Life Span," *New York Times,* March 9, 1999.

111 "Aging is not immutable.": Scott Fletcher, David Houle, and James Kurt-singer, "Aging Research Grows Up," *Genetics* 153 (October 1999): 813–23.

111 "I think it's easily possible": Austad, testimony to President's Council on Bioethics.

111 effective life-extension drugs on the market: Thomas E. Johnson, "Subfield History: *Caenorhabditis elegans* as a System for Analysis of the Genetics of Aging," *Science*'s *SAGE KE—Science of Aging Knowledge Environment* (August 28, 2002), http://sageke.sciencemag.org/cgi/content/full/sageke;2002/34/re4.

Chapter 6: Methuselah's World

114 Figure 6: The sources for this graphic are John R. Wilmoth, "Human Longevity in Historical Perspective," in Paola Timiras, ed., *Physiological Basis of Aging and Geriatrics* (Boca Raton, Fla.: CRC Publishing, 2003); Indur M. Goklany, *Economic Growth and the State of Humanity,* PERC Policy Series no. PS-21 (Bozeman, Mont.: PERC, April 2001); and "Getting Better All the Time," *The Economist,* November 8, 2001, available at http://economist.com/surveys/displaystory.cfm?story_id=841842.

115 Figure 7: The source for this graphic is the UN Department of Economic and Social Affairs, Population Division, *World Population Prospects: The 2000 Revision,* available at http://www.un.org/esa/population/publications/wpp2000/highlights.pdf.

116 one in five Americans: National Center for Chronic Disease Prevention and Health Promotion, "Healthy Aging for Older Adults," Centers for Disease Control website.

116 spend three to five times: Howard Oxley and Stephane Jacobzone, *Health Care Expenditure: A Future in Question* (Paris: Organisation for Economic Co-operation and Development, 2001).

116 Medicare costs in last year of life: See Donald R. Hoover et al., "Medical Expenditures During the Last Year of Life: Findings from the 1992–1996 Medicare Current Beneficiary Survey," *Health Services Research* 37, no. 6

(December 2002): 1625–42; estimates of cost in the last three years are based on extrapolation of the data in Lisa R. Shugarman et al., "Differences in Medicare Expenditures During the Last 3 Years of Life," *Journal of Gerontological Medicine* 19 (2004): 127–35.

117 almost half of all health-care spending: "The Demographic Facts of Life," at Population Connection website, http://www.populationconnection.org/Communications/demfacts.pdf.

117 Figure 8: Howard Oxley and Stephane Jacobzone, *Health Care Expenditure: A Future in Question* (Paris: Organisation for Economic Co-operation and Development, 2001).

117 7.5 percent of U.S. GDP in 2030: Congressional Budget Office, "A 125-Year Picture of the Federal Government's Share of the Economy, 1950 to 2075," Congressional Budget Office issue summary no. 1, revised July 3, 2003, available at http://www.cbo.gov/showdoc.cfm?index=3521&sequence=0.

120 demographic trends: See UN Department of Economic and Social Affairs, Population Division, *World Population Prospects: The 2002 Revision,* and ibid., *World Urbanization Prospects: The 2001 Revision,* http://esa.un.org/unpp.

121 "The bottom line is": S. Jay Olshansky, "Duration of Life: Is There a Biological Warranty Period?" testimony before the President's Council on Bioethics, December 12, 2002, available at http://www.bioethics.gov/transcripts/dec02/session2.html.

123 9.4 billion people: This figure results from applying the model described in the text to statistics from UN Department of Economic and Social Affairs, Population Division, *World Population Prospects: The 2002 Revision,* and ibid., *World Urbanization Prospects: The 2001 Revision.*

123 between 1970 and 1973: U.S. Bureau of the Census, International Data Base.

125 older people are more likely to vote: U.S. Bureau of the Census, http://www.census.gov/population/socdemo/voting/p20-542/tab08.xls.

Chapter 7: A Child of One's Own

128 first "test-tube baby": The story of Louise Joy Brown and the events leading up to her conception and birth are well documented; see, e.g.,

P. Gwynne, A. Collings, and B. Gastel, "The Test-Tube Baby," *Newsweek,* July 24, 1978, 76.

128 Rifkin and Kass's opposition to IVF: In *The Biotech Century: Harnessing the Gene and Remaking the World* (San Francisco: Tarcher/Putnam, 1998), Rifkin writes, "To whom, in this new era, would we entrust the authority to decide what is a good gene that should be added to the gene pool and what is a bad gene that should be eliminated?" Chris Mooney, "Irrationalist in Chief," *The American Prospect,* September 24, 2001, available at http://www.prospect.org/print/V12/17/mooney-c.html; Leon Kass, "Babies by Means of in Vitro Fertilization: Unethical Experiments on the Unborn?" *New England Journal of Medicine* 285 (1971): 1174–79.

128 more than a million babies: Joseph L. Goldstein, chairman of Lasker Awards jury, opening remarks, awards ceremony, September 2001.

128 One out of every hundred births: National Institute for Research Advancement (Japan), "The Development of Life Sciences and Law," *NIRA Policy Research* 14, no. 6 (2001), available at http://www.nira.go.jp/publ/seiken/ev14n06.html.

129 In Australia: "IVF Defects, Catalyst," Australian Broadcasting Corporation, broadcast July 17, 2003.

129 Doctors take several eggs: Aniruddha Malpani, M.D., and Anjali Malpani, M.D., *How to Have a Baby: Overcoming Infertility,* online book, http://www.infertilitybooks.com/onlinebooks/malpani/chapter25a.html (chapter 25a, "Test Tube Babies—IVF and GIFT").

129 cost of IVF: For additional information, see Minnesota Public Radio, *The Fertility Race* (series), part 7, "Twenty Years of Test-Tube Babies," http://news.mpr.org/features/199711/20_smiths_fertility/part7/section1.shtml (also available as two-CD set); National Fertility Directory, "The Cost of IVF," http://www.fertilitydirectory.org/costofivf.html; Sherman J. Silber, M.D., "Millennium Update: Future Trends in Human Infertility and Infertility Treatment," http://www.infertile.com/treatmnt/treats/millnium.htm.

129 no guarantees that a woman will get pregnant: Exactly how many embryos are flushed out of the body without implanting in the uterine wall is a matter of some debate, but it's clear that the percentage is substantial. See

M. G. R. Hull et al., "Population Study of Causes, Treatment, and Outcome of Infertility," *British Medical Journal* 291 (1985): 1693–97; A. A. Templeton and G. C. Penney, "The Incidence, Characteristics, and Prognosis of Patients Whose Infertility Is Unexplained," *Fertility and Sterility* 37 (1982): 175–82.

129 one in every six couples is infertile: Lee Silver, *Remaking Eden* (New York: Avon Books, 1997), 83.

130 the Dunthornes: Their story is told in BBC, *Superhuman* (series), "The Baby Makers," broadcast November 18, 2000; see also "Science Screens Out Defective Genes," *BBC Health Online,* broadcast November 18, 2000.

130 In PGD: Malpani and Malpani, *How to Have a Baby* (chapter 26, "PGD— Preimplantation Genetic Diagnosis—The Newest Art").

131 PGD for Huntington's disease: K. Sermon et al., "Preimplantation Genetic Diagnosis for Huntington's Disease with Exclusion Testing," *European Journal of Human Genetics* 10 (October 2002): 591–58; A. Lashwood and F. Flinter, "Clinical and Counselling Implications of Preimplantation Genetic Diagnosis for Huntington's Disease in the UK," review, *Human Fertility* 4, no. 4 (2001): 235–38.

131 several thousand PGD procedures: International Working Group on Preimplantation Genetics, "Preimplantation Genetic Diagnosis: Experience of Three Thousand Clinical Cycles," report of the eleventh Annual Meeting on Preimplantation Genetic Diagnosis, in association with the tenth International Congress of Human Genetics, Vienna, May 15, 2001, available at http://216.242.209.125/11m.shtml; A. Kuliev and Y. Verlinsky, "Current Features of Preimplantation Genetic Diagnosis," *Reproductive Biomedicine Online* 5, no. 3 (November–December 2002): 294–99.

132 the ApoE gene: For information on the gene and how its variants affect the risk of heart disease and Alzheimer's disease, see J. D. Smith, "Apolipoproteins and Aging: Emerging Mechanisms," *Ageing Research Reviews* 1, no. 3 (June 2002): 345–65.

133 ANDi: A. W. Chan et al., "Transgenic Monkeys Produced by Retroviral Gene Transfer into Mature Oocytes," *Science* 291 (January 12, 2001): 309–12.

136 total savings to society: R. A. Harris et al., "Cost Utility of Prenatal Diagnosis and the Risk-Based Threshold," *The Lancet* 363 (2004): 276–82.

136 Tay-Sachs disease: For information, see http://www.marchofdimes.com/professionals/681_1227.asp.

137 73 percent approved: Genetics and Public Policy Center, Johns Hopkins University, "Americans Deeply Divided About Use of Genetic Technologies in Reproduction," report, 2002, http://www.dnapolicy.org/research/reproductiveGenetics.jhtml?subSection=reproductive.

Chapter 8: A Child of Choice

139 the race to sequence the human genome: See, for example, J. C. Venter et al., "The Sequence of the Human Genome," *Science* 291 (February 16, 2001): 1304–51; E. Pennisi, "Academic Sequencers Challenge Celera in a Sprint to the Finish," *Science* 283 (March 19, 1999): 1822–23; Paul Jacobs and Peter G. Gosselin, "Firm's Patent Bids Alarm Gene Project Researchers," *Los Angeles Times,* October 24, 1999; Robert Langreth, "U.S., Private Gene-Sequencing Race Heats Up," *Wall Street Journal,* March 16, 1999; "The Gene Race Quickens," editorial, *Washington Post,* August 22, 1998.

141 "Starting in 1990": Eric Lander and Bob Austin, "Sequencing and Re-sequencing the Biome!" workshop summary, National Institutes of Health, National Human Genome Research Institute, July 23, 2002, available at http://www.genome.gov/10005564.

141 exponential technology improvement: Richard Dawkins, "Son of Moore's Law," in John Brockman, ed., *The Next Fifty Years* (New York: Vintage, 2002), 145–58.

143 your genome is spelled out: U.S. Department of Energy, Human Genome Program, *Genomics and Its Impact on Science and Society: A 2003 Primer,* 2003.

143 gene chips that detect at least 100,000: Malorye Branca, "Beyond the Blueprint," *Bio-IT World,* April 15, 2003; Marilee Ogren, "Whole-Genome SNP Genotyping: Four Highly Parallel Approaches Make the Technically Intractable a Reality," *The Scientist* 17, no. 11 (June 2, 2003). Online at http://www.the-scientist.com/yr2003/jun/lcprofile1_030603.html.

146 A couple goes into an infertility lab: Lee Silver, *Remaking Eden* (New York: Avon Books, 1997), 233–37.

147 only 20 percent: Johns Hopkins Genetics and Public Policy Center and

Princeton Survey Research Associates, "Public Awareness and Attitudes about Reproductive Genetic Technology," report, December 9, 2002, available at http://www.dnapolicy.org/research/reproductiveGenetics.jhtml? subSection=reproductive.

148 genes for obesity: E. E. Snyder et al., "The Human Obesity Gene Map: 2003," *Obesity Research* 12, no. 3 (March 2004): 369–439, available online at http://obesitygene.pbrc.edu/cgi-bin/ace/mainMenu.cgi.

150 "first step toward a eugenic world": Leon Kass, "How One Clone Leads to Another," *New York Times,* January 24, 2003.

152 "The person left without any choice": Bill McKibben, *Enough: Staying Human in an Engineered Age* (New York: Henry Holt, 2003), 191.

153 monoamine oxidase A: A. Caspi et al., "Role of Genotype in the Cycle of Violence in Maltreated Children," *Science* 297 (August 2, 2002): 851–54.

155 genes and environment: R. Plomin et al., *Behavioral Genetics,* 4th ed. (New York: Worth, 2000): 132.

157 we don't know Einstein's IQ: This is a hypothetical example and nothing more. For more on Einstein's mental faculties, see S. F. Witelson, D. L. Kigar, and T. Harvey, "The Exceptional Brain of Albert Einstein," *The Lancet* 353, no. 9170 (June 16, 1999): 2149–53.

158 Figure 9: Original calculation described in text.

158 genes' influence on personality: See R. Riemann, A. Angleitner, and J. Strelau, "Genetic and Environmental Influences on Personality: A Study of Twins Reared Together Using the Self- and Peer-Report NEO-FFI Scales," *Journal of Personality* 65 (1997): 449–76, and T. J. Bouchard, Jr., and J. C. Loehlin, "Genes, Evolution, and Personality," *Behavioral Genetics* 31 (2001): 243–73.

159 "Most traits are desirable at intermediate values": Stephen Pinker, "The Designer Baby Myth," *Boston Globe,* June 1, 2003.

159 COMT gene: For general information on the COMT (catechol-O-methyltransferase) gene, see the entry in the OMIM (Online Mendelian Inheritance in Man) database maintained by Johns Hopkins University, http://www.ncbi.nlm.nih.gov/entrez/dispomim.cgi?id=116790.

160 This is not to say that cloning: Margaret Talbot, "A Desire to Duplicate," *The New York Times Magazine,* February 4, 2001.

161 "psychological implications": Jeremy Rifkin and T. Howard, *Who Shall Play God?* (New York: Dell, 1977), 15.

161 "profoundly dehumanizing": Leon Kass, *Life, Liberty, and the Defense of Dignity: The Challenge of Bioethics* (San Francisco: Encounter Books, 2002), 131.

161 Other bioethicists warned of other psychological problems: F. Van Balen, "Child-Rearing Following in Vitro Fertilisation," *Journal of Child Psychology and Psychiatry* 37 (1996): 687–93.

162 conjure up a "fantasy child": L. H. Burns, "Infertility as Boundary Ambiguity: One Theoretical Perspective," *Family Processes* 26 (1997): 359–72.

162 "the children themselves": Emma Goodman, Fiona MacCallum, and Susan Golombok, "Follow-up Studies on the Psychological Consequences of Successful IVF Treatment," *Biomedical Ethics* 3, no. 2 (1998): 40–43.

163 "The notion of enhancement is deeply ingrained": Lisa M. Krieger, "Research Offers New Ways to Study, Alter Brain Function but Benefits Come with Social Policy Questions," *San Jose Mercury News*, May 21, 2002.

164 Improved nutrition is credited: J. Lawrence Angel, "Colonial to Modern Skeletal Change in the U. S. A.," *American Journal of Physical Anthropology* 45, no. 3 (November 1976): 725.

164 In 1984, political scientist James Flynn: James R. Flynn, "The Mean IQ of Americans: Massive Gains 1932 to 1978," *Psychological Bulletin* 95 (1984): 29–51; James R. Flynn, "Massive IQ Gains in 14 Nations: What IQ Tests Really Mean," *Psychological Bulletin* 101 (1987): 171–91.

166 "the impending practice of germline genetic manipulation": Kass quoted in Ralph Brave, "Governing the Genome," *The Nation*, November 21, 2001. Online at http://www.thenation.com/dec.mhtml?i-20011210d5-brave.

166 "modern genetics is eugenics": Annas quoted in John Cavanaugh-O'Keefe, "The Roots of Racism and Abortion: An Exploration of Eugenics," online book, http://www.eugenics-watch.com/roots/index.html.

166 "Our present condition is this": Gilbert Meilaender, "Designing Our Descendants," *First Things* 109 (January 2001): 25–28.

169 two thirds of the couples: L. Hennen and A. Sauter, "Pre-implantation Genetic Diagnosis (PGD)—Regulation and Medical Practice," paper presented at European Parliamentary Technology Assessment Network (EPTA) annual conference, "Research Involving Human Beings," Bern,

Switzerland, October 27–28, 2003, available at http://www.eptanetwork. org/EPTA/documents/2003/031027_TAB.pdf.

Chapter 9: The Wired Brain

172 Johnny Ray: Johnny Ray, "Me, Myself, My Implants, My Micro-processors, and I," *Software Development Magazine* (September 2000), available at http://www.sdmagazine.com/documents/sdm0009a/.

172 Every day in the United States: For further information on stroke, see American Stroke Association, "About Stroke," http://www.strokeassociation.org/ presenter.jhtml?identifier=11402 (statistics on stroke frequency); University of Alabama, National SCI (Spinal Cord Injury) Statistical Center, "Spinal Cord Injury: Facts and Figures at a Glance" (Birmingham: University of Alabama at Birmingham Press, May 2001); see also the center's website, http://www.spinalcord.uab.edu/show.asp?durki=21446; Nell Boyce, "Enter the Cyborgs," *U.S. News & World Report,* May 13, 2002: 56–58.

173 Phil Kennedy: Author interview with Phil Kennedy at National Institutes of Health Neural Prosthetics Workshop, 2002.

173 Bakay implanted the device: P. R. Kennedy et al., "Direct Control of a Computer from the Human Central Nervous System," *IEEE Transactions on Rehabilitation Engineering* 8, no. 2 (June 2000): 198–202.

174 brain-computer interface: M. Nicolelis, "Brain-Machine Interfaces to Restore Motor Function and Probe Neural Circuits," *Nature Reviews Neuroscience* 4 (May 2003): 417–22.

177 Nicolelis: Carl Zimmer, "Mind over Machine," *Popular Science* (February 2004): 48–53; for a biography of Nicolelis, see http://www.nicolelislab.net.

178 rat could control a robot arm: J. K. Chapin et al., "Real-Time Control of a Robot Arm Using Simultaneously Recorded Neurons in the Motor Cortex," *Nature Neuroscience* 2, no. 7 (July 1999): 664–70; J. Wessberg et al., "Real-Time Prediction of Hand Trajectory by Ensembles of Cortical Neurons in Primates," *Nature* 408 (November 16, 2000): 361–65.

178 Their first breakthrough: J. M. Carmena et al., "Learning to Control a Brain-Machine Interface for Reaching and Grasping by Primates," *PLoS Biology* 1, no. 2 (November 2003): E42.

180 "The monkey suddenly realized": Duke University Medical School, "Monkeys Consciously Control a Robot Arm Using Only Brain Signals; Appear to 'Assimilate' Arm As If It Were Their Own," DukeMed News Release, October 13, 2003.

181 In 2004, the researchers tested: Maggie Fox, "Thought-Controlled Arm May Work in People," report, Reuters Science, March 23, 2004.

182 first cochlear implants: K. I. Berliner and W. F. House, "Cochlear Implants: An Overview and Bibliography," review, *American Journal of Otology* 2, no. 3 (January 1981): 277–82.

182 There are 34 million deaf: D. L. Blackwell, J. G. Collins, and R. Coles, "Summary Health Statistics for U.S. Adults: National Health Interview Survey, 1997," *Vital Health Statistics* 10 (2002): 205.

182 Today, more than seventy thousand patients worldwide: National Institutes of Health data, http://www.nidcd.nih.gov/health/hearing/coch.asp#c.

182 auditory nerve has more than thirty thousand fibers: Robert Finn et al., "Sound from Silence: The Development of Cochlear Implants," *Beyond Discovery* (online anthology), National Academy of Sciences, 1998, see http://www.beyonddiscovery.org/content/view.article.asp?a=252 for a summary and link to article.

183 research is ongoing into implants: P. J. Rousche and R. A. Normann, "Chronic Intracortical Microstimulation (ICMS) of Cat Sensory Cortex Using the Utah Intracortical Electrode Array," *IEEE Transactions of Rehabilation Engineering* 7, no. 1 (March 1999): 56–68.

183 In September 2002: Steven Kotler, "Vision Quest," *Wired,* September 2002, available at http://www.wired.com/wired/archive/10.09/vision.html.

185 The focus of visual prosthetic research: W. H. Dobelle, "Artificial Vision for the Blind by Connecting a Television Camera to the Visual Cortex," *ASAIO (American Society for Artificial Internal Organs) Journal* 46, no. 1 (January–February 2000): 3–9.

187 Garrett Stanley's experiment: Garrett B. Stanley, Fei F. Li, and Yang Dan, "Reconstruction of Natural Scenes from Ensemble Responses in the Lateral Geniculate Nucleus," *Journal of Neuroscience* 19, no. 18 (September 15, 1999): 8036–42.

188 communicating sounds directly from brain to brain: H. Hoag, "Neuroengineering: Remote Control," *Nature* 423 (June 19, 2003): 796–98.

189 *sensation equals recollection equals imagination*: Stephen M. Kosslyn, Giorgio Ganis, and William L. Thompson, "Neural Foundations of Imagery," *Nature Reviews Neuroscience* 2 (2001): 635–42.

190 There is, as yet, no cure for Parkinson's: P. Tuite and J. Riss, "Recent Developments in the Pharmacological Treatment of Parkinson's Disease," *Expert Opinion on Investigational Drugs* 12, no. 8 (August 2003): 1335–52.

190 When he accidentally stimulated the thalamus: B. Pollak et al., "Treatment of Parkinson's Disease: New Surgical Treatment Strategies," *European Neurology* 36, no. 6 (1996): 400–4.

191 deep-brain stimulators: For recent developments, see A. L. Benabid et al., "Long-Term Electrical Inhibition of Deep Brain Targets in Movement Disorders," *Movement Disorders* 13, Suppl. 3 (1998): 119–25; F. Clancy, "Following the Leads," *Minnesota Medicine* 85 (April 2002), available at http://www.mnmed.org/publications/MNMed2002/April/Clancy.html.

191 patients with severe obsessive-compulsive disorder: B. D. Greenberg and A. R. Rezai, "Mechanisms and the Current State of Deep Brain Stimulation in Neuropsychiatry," review, *CNS Spectrums* 8, no. 7 (July 2003): 522–26.

191 additional evidence that electrical stimulation: T. Satow et al., "Mirth and Laughter Arising from Human Temporal Cortex," *Journal of Neurology, Neurosurgery, and Psychiatry* 74, no. 7 (July 2003): 1004–5.

192 Delgado's experiments: J. M. R. Delgado, *Physical Control of the Mind* (New York: Harper & Row, 1969).

196 more than 5 million people: Centers for Disease Control, National Center for Injury Prevention and Control (NCIPC), "Traumatic Brain Injury Facts," at CDC website, http://www.cdc.gov/doc.do?id=0900f3ec800081d7; D. J. Thurman et al., "Traumatic Brain Injury in the United States: A Public Health Perspective," *Journal of Head Trauma Rehabilitation* 14, no. 6 (December 1999): 602–15.

197 Theodore Berger: Ted Berger's work is described in Gregory Huang, "Mind-Machine Merger," *MIT Technology Review* (May 2003): 38–45.

199 DiCarlo's experiments: Discussed in ibid.

Chapter 10: World Wide Mind

208 implants fabricated like microchips: For an overview of this research, see K. D. Wise, http://www.eecs.umich.edu/~wise/Research/Overview/Wise_Research.pdf.

208 German firm Infineon: Infineon, "Neuro-Chip from Infineon Can Read Your Mind—New Findings in Brain Research Expected," news release, November 2, 2003, at http://www.infineon.com/cgi/ecrm.dll/jsp/showfrontend.do?lang=EN&content_type=NEWS&content_oid=57550&news_nav_oid=-9979.

209 overcoming obstacles to brain surgery: Eskandar et al., "Surgery for Parkinson Disease in the United States, 1996 to 2000," on deep-brain stimulators, *Journal of Neurosurgery* 99 (2003): 863.

210 electroencephalography: I. Wickelgren, "Power to the Paralyzed," *Science* 299 (January 24, 2003): 497, and Wickelgren, "Tapping the Mind," *Science* 299 (January 24, 2003): 496–99.

211 limitations of functional magnetic resonance imaging: S. Saini et al., "Technical Cost of Radiologic Examinations: Analysis Across Imaging Modalities," *Radiology* 216, no. 1 (July 2000): 269–72; S.-G. Kim et al., "Spatial Resolution of BOLD and Other Functional MRI Techniques," in C. Moonen and P. A. Bandettini, eds., *Medical Radiology—Diagnostic Imaging and Radiation Oncology, Functional MRI* (New York: Springer-Verlag, 1999), 195–203.

212 transcranial magnetic stimulation (TMS): B. Kloza, "Watching Living Brains," *ScienCentralNEWS* (online periodical), January 21, 2003, available at http://www.sciencentral.com/articles/view.php3?language=english&type=&article_id=218391866; A. Cowey and V. Walsh, "Tickling the Brain: Studying Visual Sensation, Perception and Cognition by Transcranial Magnetic Stimulation," review, *Progress in Brain Research* 134 (2001): 411–25.

213 10 million people: "LASIK and Its Alternatives: An Update," *Medical Letter on Drugs and Therapeutics* 46, no. 1174 (January 19, 2004): 5–7

213 endovascular surgery for the brain: A. S. Boulos et al., "Endovascular Aneurysm Treatment: A Proven Therapy," review, *Neurological Research* 24,

Suppl. 1 (2002): S71–79; P. Bairstow et al., "Comparison of Cost and Out-come of Endovascular and Neurosurgical Procedures in the Treatment of Ruptured Intracranial Aneurysms," *Australasian Radiology* 4a6, no. 3 (September 2002): 249–51.

214 125,000 endovascular treatments: J. Rosenorn, V. Eskesen, and K. Schmidt, "Unruptured Intracranial Aneurysms: An Assessment of the Annual Risk of Rupture Based on Epidemiological and Clinical Data," review, *British Journal of Neurosurgery* 2, no. 3 (1988): 369–77.

217 "Our analyses of the brain": Nicolelis, "Monkeys Move Matter, Mentally," *Duke Magazine* (January–February 2004). Online at http://www.dukemagazine.duke.edu/dukemag/issues/010204/depgaz8.html.

222 While undermining central authority, information technology increases: Robert Wright, *Nonzero: The Logic of Human Destiny* (New York: Vintage, 2001).

Chapter 11: Life Without Limits

227 "We are snipping the very last weight": Bill McKibben, *Enough: Staying Human in an Engineered Age* (New York: Henry Holt, 2003), 47.

227 "The human soul yearns": Leon Kass, *Life, Liberty, and the Pursuit of Dignity: The Challenge of Bioethics* (San Francisco: Encounter Books, 2002), 269.

230 our chimp ancestor's brain: H. J. Jerison, *Evolution of the Brain and Intelligence* (New York: Academic Press, 1973), 87.

230 Figure 10: H. J. Jerison, *Evolution of the Brain and Intelligence* (New York: Academic Press, 1973).

Index

About the Author

RAMEZ NAAM helped develop two of the most widely used pieces of software in the world—Microsoft Internet Explorer and Microsoft Outlook. He serves as an advisor to several technology associations and speaks frequently at technology and futures conferences. He lives in Seattle.